High Speed Pulse and Digital Techniques

HIGH SPEED PULSE AND DIGITAL TECHNIQUES

Arpad Barna
Hewlett-Packard Laboratories
Palo Alto, California

A Wiley-Interscience Publication

JOHN WILEY & SONS, New York   •   Chichester   •   Brisbane   •   Toronto

Library of Congress Cataloging in Publication Data:

Barna, Arpad.
  High speed pulse and digital techniques.

  "A Wiley-Interscience publication."
  Bibliography: p.
  Includes index.
  1. Digital integrated circuits. 2. Pulse
circuits. 3. Transistor circuits. I. Title.

TK7874.B38   621.381'73   79-26264
ISBN 0-471-06062-3

Printed in the United States of America

10 9 8 7 6 5 4 3 2 1

To California

PREFACE

The rapid advances in semiconductor technology have led to an increasing variety, improving performance, decreasing cost, and expanding application of high speed digital integrated circuits.

In addition to conventional circuit design, the efficient utilization of these circuits also requires the use of high speed pulse and digital techniques that up to now could be found only scattered among many books on circuit theory, pulse circuits, and computer-aided circuit design, and in various catalogs and journals. This book presents such techniques in one volume and in a form oriented toward the user of high speed digital circuits. It is based on the author's 25 years of experience in high speed pulse and digital techniques.

A complete treatment of the subject requires the use of calculus and complex variables. Nevertheless, a prior knowledge in these fields is not needed for this book, except for footnotes and some optional problems. However, some elementary features of calculus are introduced and used in the text. The presentation is liberally interspersed with worked examples that support the introduction of new concepts. The problems at the end of the chapters enable the reader to broaden and test his understanding of the material; answers to selected problems are given at the end of the book.

ARPAD BARNA

Palo Alto, California
December 1979

CONTENTS

†Optional material is denoted by † .

High Speed Pulse and Digital Techniques

CHAPTER 1

OVERVIEW

This book describes pulse and digital techniques that are applicable to the use of today's high speed digital integrated circuits with operating speeds of 1 to 10 nanoseconds. Perhaps not too surprisingly the use of such circuits presents more problems in interconnections and the associated time constraints than slower circuitry. For this reason a substantial part of the book deals with passive R-C, R-L, R-L-C, and transmission line circuits. However, many of these considerations are strongly related to the internal structure of the integrated circuits, hence a treatment of the high speed properties of diodes and bipolar transistors is also included. Further, properties of the two fastest digital integrated circuit families, the ECL and the Schottky-diode-clamped TTL, are also discussed.

Chapter 2 reviews basic results of linear circuit theory. However, in addition to the traditional treatment emphasizing wideband amplifiers, the chapter also presents material that is oriented toward problems arising from grounding and crosstalk in digital systems.

Chapter 3 provides a treatment of diode circuits. Properties of junction diodes are discussed in detail, as they are basic to the understanding of transistor operation. Computer-aided design methods are introduced and applied to dc and transient analysis in diode circuits—these methods are also applicable to transistor circuits. A brief discussion on tunnel diodes and tunnel-diode circuits is also included.

Chapter 4 treats bipolar (junction) transistors and circuits using them. A complete description of a high speed bipolar transistor would require a model using about 26 parameters. However, the treatment here is restricted to simple models focusing on properties that are of principal importance in high speed pulse and digital circuits. The chapter also includes a discussion on emitter

1

follower stability, a description of Schottky-diode-clamped TTL circuits, and an analysis of propagation delays and transition times in emitter-coupled logic (ECL) circuits.

Chapter 5 provides a description of transmission lines and their use in digital systems. It treats propagation delay, capacitance, and inductance, and describes coaxial, stripline, and various other transmission line configurations that are used in digital systems. Transients in transmission lines are analyzed for linear resistive and capacitive terminations, as well as for nonlinear resistive termination encountered in use with TTL circuits. A brief discussion of losses in transmission lines is also included.

CHAPTER 2

LINEAR CIRCUITS

This chapter describes basic properties of linear components and circuits with emphasis on characteristics that are utilized in high speed digital circuits. Time-domain properties of resistors, capacitors, inductors, and voltage and current sources are described, followed by the introduction of the unit step, the exponential, and the logarithmic functions.

Transient responses of R-C circuits are described for step, pulse, and ramp inputs, and the Elmore delay and the Elmore risetime are introduced as characteristics of the frequency response. Transient responses of R-L-C circuits are presented including applications to crosstalk on ground returns; delay, risetime, overshoot, and frequency response are also discussed, and relationships are established between the transient response and the Elmore delay and the Elmore risetime.

The chapter also includes brief discussions of R-L circuits and pulse transformers and concludes by a treatment of cascaded R-C and R-L-C circuits. Transmission lines are not discussed, since they are the subject of Chapter 5.

2.1 RESISTORS

The simplest linear component is the *resistor*,* shown in Figure 2.1a. When the applied *voltage* V_R (measured in volts, V) and the *resistance R* (measured in ohms, Ω) are given, the *current I_R* (measured in amperes, A) can be found from *Ohm's law*:

*Terms are introduced by italics.

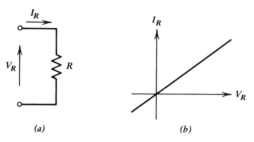

Figure 2.1   The resistor. (*a*) Symbol; (*b*) I_R versus V_R characteristic.

$$I_R = \frac{V_R}{R} \qquad (2.1)$$

illustrated in Figure 2.1*b*.

The *power P_R* (measured in watts, W) dissipated in a resistor is

$$P_R = V_R I_R \qquad (2.2a)$$

which can be also written as

$$P_R = I_R^2 R, \qquad (2.2b)$$

or as

$$P_R = \frac{V_R^2}{R}. \qquad (2.2c)$$

The resistance of a bar of material is given by

$$R = \rho \frac{l}{A} \qquad (2.3)$$

where ρ is the *resistivity* (for copper $\rho \approx 1.7 \times 10^{-8}$ Ωm), l is the length of the bar in meters, and A is its cross-sectional area in square meters (m²).

Example 2.1   Calculate the resistance of 100 feet of #20 copper wire. The length is $l = 100$ feet ≈ 30 m. The diameter is given by wire tables as 0.032 in. = 0.81 mm = 0.81×10^{-3} m. Thus the cross-sectional area

$$A = 0.81^2 \times 10^{-6}\,\mathrm{m}^2\,\frac{\pi}{4} = 0.52 \times 10^{-6}\,\mathrm{m}^2.$$

Hence

$$R = \rho \frac{l}{A} = 1.7 \times 10^{-8}\ \Omega\mathrm{m}\,\frac{30\ \mathrm{m}}{0.52 \times 10^{-6}\,\mathrm{m}^2} = 1.04\ \Omega.$$

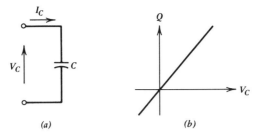

Figure 2.2 The capacitor. (*a*) Symbol; (*b*) Q versus V_C characteristic.

2.2 CAPACITORS

A *capacitor* (Figure 2.2*a*) is capable of storing *charge*. For a given voltage V_C and *capacitance* C (measured in farads, F: 1 farad = 1 second/ohm), the stored charge Q (measured in coulombs: 1 coulomb = 1 ampere $\times$ second) is given by

$$Q = CV_C \qquad (2.4)$$

as illustrated in Figure 2.2*b*. The stored charge is, however, the accumulation of current. Hence, when the current is I_C, in a time interval with a duration dt the charge changes by an amount dQ given by

$$dQ = I_C \, dt; \qquad (2.5a)$$

also, the current I_C equals the *rate of change* of Q, which is dQ/dt:

$$I_C = \frac{dQ}{dt}. \qquad (2.5b)$$

For a voltage change of dV, according to eq. (2.4), the charge changes by an amount of

$$dQ = C \, dV_C. \qquad (2.6)$$

Combination of eqs. (2.5*b*) and (2.6) leads to

$$I_C = C \frac{dV_C}{dt}. \qquad (2.7)$$

Example 2.2 A capacitor that has a capacitance of $C = 0.5$ farad is connected to a current source that delivers the current I_C shown in the upper graph of Figure 2.3. The charge Q of the capacitor and the voltage V_C across it are shown in the lower graph for an arbitrarily chosen initial charge of $Q_{t=0} = 0.25$ coulomb. The quantities I_C, Q, and V_C are related by eqs. (2.4) through (2.7).

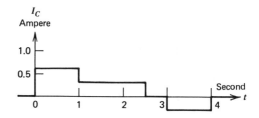

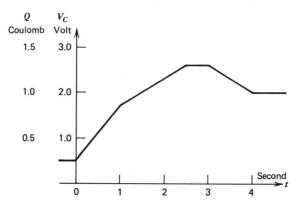

Figure 2.3 Current I_C flowing into a capacitor with a capacitance of 0.5 farad (upper graph), and the voltage V_C and the charge Q of the capacitor with an arbitrary initial charge of 0.25 coulomb (lower graph).

In Figure 2.3 the graph of the current flowing into the capacitor is composed of constant-current segments. As a result, the charge accumulated in a time interval during which the current is constant can be computed as the current multiplied by the duration of the time interval. In general, the change of charge is given by the area under the graph of the current; the area is counted negative when the current is negative.

Example 2.3 A current I_C flowing into a capacitor and the resulting charge are shown in Figure 2.4 where we assumed zero initial charge. The area under the graph of the current during the time interval of $t = 0$ to $t = 2$ seconds is given by the area of the triangle as $I_C t/2$; thus, for example, *at* $t = 1$ second the area is 0.5 ampere $\times$ 1 second/2 = 0.25 ampere $\times$ second = 0.25 coulomb, as shown in the graph of Q. The charge can be found in a similar manner at any time between $t = 0$ and $t = 2$ seconds.

Between $t = 2$ seconds and $t = 3$ seconds $I_C = 0$, hence Q remains unchanged. Between times $t = 3$ seconds and $t = 5$ seconds $I_C = -1$ ampere, thus Q changes linearly. The charge accumulated during this interval is $dQ = -1$ ampere $\times$ (5 seconds $-$ 3 seconds) $= -2$ coulombs resulting in a change from the $Q = 1$ coulomb at $t = 3$ seconds to $Q = -1$ coulomb *at* $t = 5$ seconds.

Between $t = 5$ seconds and $t = 6$ seconds $I_C = 0$, thus Q remains -1 coulomb. Between $t = 6$ seconds and, for example, $t = 7$ seconds the change of charge is given by the area of the trapezoid: (7 seconds $-$ 6 seconds) $\times$ (1 ampere $+$ 0.5 ampere)/2 $=$ 0.75 coulomb. By adding this to the -1 coulomb charge that is present at $t = 6$ seconds, we get a $Q = -0.25$ coulomb at $t = 7$ seconds. The charge can be similarly found at any time between $t = 6$ seconds and $t = 8$ seconds.

Note that Figure 2.4 does not assume any specific value of capacitance C, and the voltage V_C across the capacitor is not given. However, once the charge Q as function of time is determined, the voltage as function of time can be found by use of eq. (2.4) as $V_C = Q/C$.

The charge Q in Figure 2.4 is computed as the area under the graph of I_C by use of formulae for the areas of the triangle and the trapezoid. This procedure is applicable whenever such a formula is available for the given current

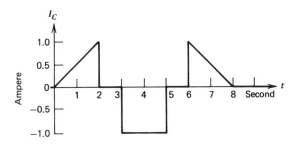

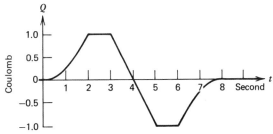

Figure 2.4 Current I_C flowing into a capacitor (upper graph) and the charge Q of the capacitor with zero initial charge (lower graph).

shape. When this is not the case the area can still be found by counting the number of squares under the graph of I_C plotted on a graph paper with a square grid.[†]

Thus far we assumed that the current I_C as function of time was given, and we sought charge Q and voltage V_C as functions of time. When the opposite holds, current I_C can be found as the rate of change of charge Q.

Example 2.4 In the lower graph of Figure 2.3, between $t = 1$ second and $t = 2.5$ seconds the charge changes by $dQ = 1.3$ coulomb – 0.85 coulomb = 0.45 coulomb. According to eq. (2.5b), the current is given as the change in charge, dQ, divided by the duration of the time interval which is $dt = 2.5$ seconds – 1 second = 1.5 second. Thus, $I_C = dQ/dt = 0.45$ coulomb/1.5 second = 0.3 ampere, in agreement with the upper graph of Figure 2.3.

In general, the rate of change *at* any time t can be found as the slope of the tangent drawn to the graph of Q at time t–in agreement with Figure 2.4.[‡]

A capacitor can not dissipate power, however, it can store energy. The energy E (measured in joules, J : 1 joule = 1 watt $\times$ second) stored in a capacitor with a capacitance of C is given as

$$E = \tfrac{1}{2} CV_C^2 \qquad (2.8a)$$

or as

$$E = \tfrac{1}{2} QV_C. \qquad (2.8b)$$

The capacitance of two parallel plates that have opposing areas of A each and that are separated by a distance d can be approximated as

$$C = \epsilon_0 \epsilon_r \frac{A}{d}. \qquad (2.9)$$

In eq. (2.9), $\epsilon_0 = 8.85 \times 10^{-12}$ farad/m, ϵ_r is the relative *dielectric constant* ($\epsilon_r \approx$ 1 in vacuum or in air); it is also assumed that each dimension of A is much larger than d.

Example 2.5 Calculate the capacitance between two opposing dimes spaced $d = 0.1$ in. = 2.54 mm apart in air. The diameter of a dime is 0.7 in. = 17.5 mm = 1.75×10^{-2} m. The area is thus $(1.75 \times 10^{-2}$ m$)^2 \times \pi/4 = 2.4 \times 10^{-4}$ m^2. The capacitance

---

[†]This is, in fact, a graphical integration. In general, the charge $Q = CV_C$ is given by the integral $\int I_C \, dt$ and graphical (or numerical) integration can be avoided whenever $\int I_C \, dt$ is available from a table of integrals.

[‡]Analytically, the current is given by the derivatives $I_C = dQ/dt = C \, dV_C/dt$.

$$C = \epsilon_0 \epsilon_r \frac{A}{d} = 8.85 \times 10^{-12} \frac{F}{m} \frac{2.4 \times 10^{-4} m^2}{2.54 \times 10^{-3} m}$$

$$= 0.84 \times 10^{-12} \ F = 0.84 \ pF \ (picofarad).$$

2.3 INDUCTORS

An *inductor* (Figure 2.5a) is capable of storing magnetic *flux*. For a given current I_L and *inductance* L (measured in henrys, H: 1 henry = 1 ohm $\times$ second), the stored *flux*, Φ (measured in webers: 1 weber = 1 volt $\times$ second) is given by

$$\Phi = LI_L \tag{2.10}$$

as illustrated in Figure 2.5b. When the voltage across the inductor is V_L, in a time interval with a duration of dt its flux changes by an amount $d\Phi$ given as

$$d\Phi = V_L \ dt; \tag{2.11a}$$

also, the voltage V_L equals the rate of change of the flux, which is $d\Phi/dt$:

$$V_L = \frac{d\Phi}{dt}. \tag{2.11b}$$

For a current change of dI_L, according to eq. (2.10), the flux changes by an amount of

$$d\Phi = L \ dI_L. \tag{2.12}$$

Combination of eqs. (2.11b) and (2.12) leads to

$$V_L = L \frac{dI_L}{dt}. \tag{2.13}$$

An inductor can not dissipate power, but it can store energy. The energy E stored in an inductor with an inductance of L is given as

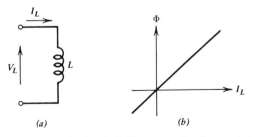

(a)   (b)

Figure 2.5 The inductor. (*a*) Symbol; (*b*) Φ versus I_L characteristic.

$$E = \tfrac{1}{2} LI_L^2 \qquad\qquad\qquad (2.14a)$$

or as

$$E = \tfrac{1}{2} \Phi I_L . \qquad\qquad\qquad (2.14b)$$

Note that the equations pertaining to inductors are similar to those of capacitors, thus the methods of the preceding section on capacitors are also applicable to inductors.[†]

2.4 VOLTAGE SOURCES AND CURRENT SOURCES

Figure 2.6a shows the symbol of a *voltage source* (or *voltage generator*). Figure 2.6b its current versus voltage characteristic. For any *finite* current I_G, the voltage source delivers a voltage $V_G = V_0$, which can be a function of time and can be *controlled* by another voltage or by a current. Figure 2.7a and Figure 2.7b show two commonly used symbols of a *current source* (or *current generator*), Figure 2.7c the current versus voltage characteristic. For any *finite* voltage

[†]Analytically the flux $\Phi = LI$ is given by the integral $\int V\, dt$, and the voltage is given by the derivatives $V = d\Phi/dt = L\, dI/dt$.

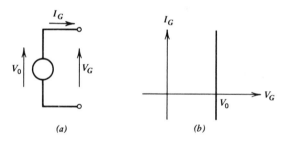

(a)   (b)

Figure 2.6 Voltage source. (a) Symbol; (b) current versus voltage characteristic.

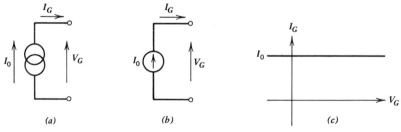

(a)   (b)   (c)

Figure 2.7 Current source. (a), (b) Two commonly used symbols; (c) current versus voltage characteristic.

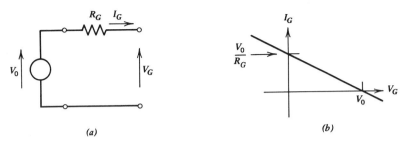

Figure 2.8 Voltage source V_0 with a source resistance R_G. (*a*) Configuration; (*b*) current versus voltage characteristic.

V_G, the current source delivers a current $I_G = I_0$, which can be a function of time and can be controlled by another current or by a voltage.

The concepts of voltage source and current source are idealizations—this is, of course, also the case for the resistor, capacitor, and inductor described in the preceding sections. The power delivered by either a voltage source or a current source is $V_G I_G$, which would increase without limit as either the current I_G drawn from the voltage source or the voltage V_G appearing across the current source increases. The simplest limiting mechanism is a resistor in series with the voltage source as shown in Figure 2.8, or a resistor in parallel with the current source as shown in Figure 2.9. The circuits of Figures 2.8*a* and 2.9*a* are equivalent and indistinguishable if $V_0 = R_G I_0$ —however, such a simple equivalence is possible only when R_G is a pure resistance.

Figure 2.10 shows a voltage source with a source resistance R_G, loaded by a load resistance R_L. In this circuit voltage V_L is given by

$$V_L = V_0 \frac{R_L}{R_L + R_G}. \tag{2.15}$$

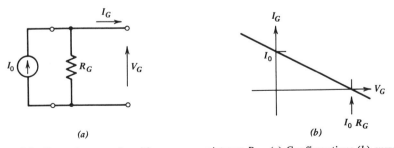

Figure 2.9 Current source I_0 with a source resistance R_G. (*a*) Configuration; (*b*) current versus voltage characteristic.

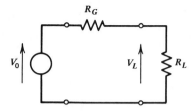

Figure 2.10 Voltage source with a source resistance R_G loaded by a load resistance R_L.

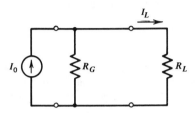

Figure 2.11 Current source with a source resistance R_G loaded by a load resistance R_L.

Figure 2.11 shows a current source with a source resistance R_G, loaded by a load resistance R_L. In this circuit current I_L is given by

$$I_L = I_0 \frac{R_G}{R_G + R_L}. \tag{2.16}$$

2.5 THE UNIT STEP, THE EXPONENTIAL, AND THE LOGARITHMIC FUNCTIONS

The upper graph in Figure 2.3 shows a current signal as function of time. However, we have yet no means to describe such a signal by an equation. The basic function for such a purpose is the *unit step function* $u(t)$ shown in Figure 2.12. We can see that

$$u(t) = 0 \text{ for } t < 0,$$
$$u(t) = 1 \text{ for } t > 0,$$
$$u(t) \text{ is undefined for } t = 0. \tag{2.17}$$

Several basic properties of the unit step function $u(t)$ can be established by inspection. (1): $u(at) = u(t)$ for any $a > 0$; (2): $[u(t)]^2 = u(t)$; (3): $u(-t) = 1 - u(t)$; (4): $u(t - t_1) = 0$ for $t < t_1$, $u(t - t_1) = 1$ for $t > t_1$, and $u(t - t_1)$ is undefined for $t = t_1$; (5): a step function of height A can be generated from the unit step function as $Au(t)$; thus $Au(t) = 0$ for $t < 0$, $Au(t) = A$ for $t > 0$, and

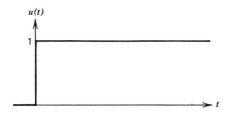

Figure 2.12 The unit step function.

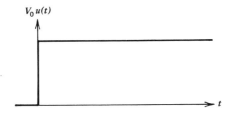

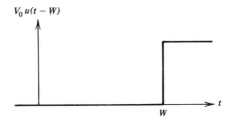

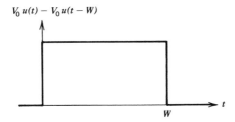

Figure 2.13 A pulse with a width of W described by the superposition of two step functions.

$Au(t)$ is undefined for $t = 0$; (6): The superposition of two step functions describes a *pulse* as shown in Figure 2.13.

The step function may be also used to describe more complex signals.

Example 2.6 A current signal $I(t)$ is given in the upper graph of Figure 2.3. It can be described by use of step functions as

$$I_C(t) = 0.6u(t) - 0.3u(t - 1) - 0.3u(t - 2.5)$$
$$- 0.3u(t - 3) + 0.3u(t - 4),$$

where the dimensions ampere and second have been omitted.

The *exponential function* may be defined as a function $y(x)$ that has a rate of change dy/dx that is equal to the function itself:

$$\frac{dy}{dx} = y. \tag{2.18}$$

Let us denote the value of y at $x = 0$ by A:

$$y(x = 0) = A. \tag{2.19}$$

From eqs. (2.18) and (2.19) we get for the rate of change of y at $x = 0$:

$$\left[\frac{dy}{dx}\right]_{x=0} = A. \tag{2.20}$$

By use of eqs. (2.19) and (2.20) we can approximate the value of y for a "small" $x = \Delta x$ as

$$y(x = \Delta x) \approx y(x = 0) + \left[\frac{dy}{dx}\right]_{x=0} \Delta x = A + A \Delta x = A(1 + \Delta x). \tag{2.21}$$

Similarly,

$$y(x = 2\Delta x) \approx y(x = \Delta x) + \left[\frac{dy}{dx}\right]_{x=\Delta x} \Delta x$$
$$= A(1 + \Delta x) + A(1 + \Delta x)\Delta x = A(1 + \Delta x)^2. \tag{2.22}$$

This procedure can be extended to an arbitrary $x = n\, \Delta x$ as

$$y(x = n\, \Delta x) \approx A(1 + \Delta x)^n, \tag{2.23a}$$

that is,

$$y(x = n\, \Delta x) \approx A\left(1 + \frac{x}{n}\right)^n. \tag{2.23b}$$

For a *given* x the approximation of eq. (2.23b) becomes more accurate as n is increased; in the limit as $\Delta x \to 0$, hence as $n \to \infty$, for $A = 1$ eq. (2.23b) becomes

$$y(x) = \lim_{n \to \infty} y(x = n\,\Delta x) = \lim_{n \to \infty} \left(1 + \frac{x}{n}\right)^n. \qquad (2.24)$$

It can be shown that the limit exists: the expression on the rightmost side of eq. (2.24) is designated the exponential function e^x. It can be also shown that an infinite series for e^x is given as

$$e^x = \sum_{k=0}^{\infty} \frac{x^k}{k!} = 1 + x + \frac{x^2}{2} + \frac{x^3}{6} + \cdots. \qquad (2.25)$$

Thus the value of $e = 1 + 1 + \frac{1}{2} + \frac{1}{6} + \cdots \approx 2.71$. The exponential function e^x, often also denoted as $\exp(x)$, is shown in Figure 2.14.

Some properties of the exponential function e^x are the following. (1): The rates of change $d(e^x)/dx = e^x$ and $d(e^{-x})/dx = -e^{-x}$. (2): $e^{a+b} = e^a e^b$ and $e^{a-b} = e^a/e^b$. (3): $e^{ab} = (e^a)^b = (e^b)^a$. Additional properties involve the logarithmic function and are described under that function.

The *logarithmic function* is the inverse of the exponential function:

$$y = \ln x \qquad (2.26)$$

when $x = e^y$. The logarithmic function $\ln x$, often also denoted as $\log_e x$, is

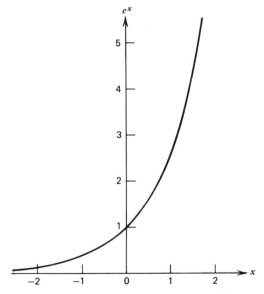

Figure 2.14 The exponential function e^x.

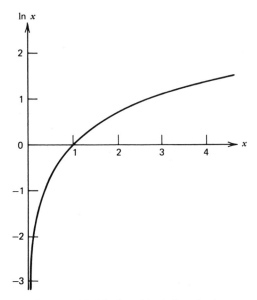

Figure 2.15   The logarithmic function $\ln x$.

shown in Figure 2.15. Some properties of the logarithmic function are the following. (1): $\ln (ab) = \ln a + \ln b$; (2): $\ln (a/b) = \ln a - \ln b$; (3): $\ln (a^b) = b \ln a$; (4): $\ln a = \log_e a = (\log_{10} a) \times (\ln 10) = (\log_{10} a)/(\log_{10} e)$; (5): $\ln a = \log_e a = (\log_b a) \times (\ln b) = (\log_b a)/(\log_b e)$ for any b; (6): $a^x = e^{x \ln a}$ for any a.

The logarithmic function can be expanded into several different series, however, these converge comparatively slowly. One such series is

$$\ln (1 + x) = \sum_{k=1}^{\infty} (-1)^{k-1} \frac{x^k}{k} = x - \frac{x^2}{2} + \frac{x^3}{3} \mp \cdots \qquad (2.27a)$$

valid when

$$-1 < x < 1. \qquad (2.27b)$$

2.6   R-C CIRCUITS

This section describes the properties of R–C circuits consisting of a single resistor and a single capacitor. Specifically, we derive voltage V_C and V_R as functions of time for various input signals V_{in} in the circuit of Figure 2.16a.

When $I_{in} = V_{in}/R$ and the values of the R's and C's are identical in Figures 2.16a and 2.16b, capacitor voltage V_C in Figure 2.16a equals capacitor voltage

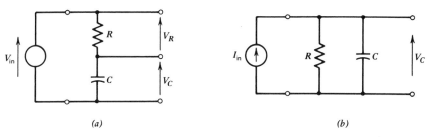

Figure 2.16 Two forms of an R-C circuit that are equivalent if $I_{in} = V_{in}/R$.

V_C in Figure 2.16*b*; however, unlike in the circuit of Figure 2.16*a*, in Figure 2.16*b* the voltage across resistor R equals V_C, and the current flowing through resistor R equals V_C/R.

2.6.1 Step Voltage Input

The simplest case is when, as shown in the upper graph of Figure 2.17, $V_{in} = V_0 u(t)$. For simplicity we shall also assume—at least for the time being—that the initial value of V_C, $(V_C)_{t \leqslant 0} = 0$.

The voltage across resistor R in Figure 2.16*a* is

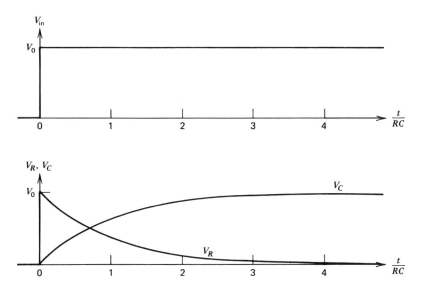

Figure 2.17 Signals in the R-C circuit of Figure 2.16*a*: Input voltage V_{in} (upper graph), and voltages V_R and V_C for $(V_C)_{t \leqslant 0} = 0$ (lower graph).

$$V_R = V_{in} - V_C.$$ (2.28)

Since there is no other place to go, the current I flowing through the resistor flows into the capacitor. This current has a value of

$$I = \frac{V_R}{R}.$$ (2.29)

Also, V_{in} is constant for $t > 0$, hence the rate of change of V_C, dV_C/dt, can be written by use of eq. (2.28) as

$$\frac{dV_C}{dt} = \frac{d(V_{in} - V_R)}{dt} = \frac{dV_{in}}{dt} - \frac{dV_R}{dt} = -\frac{dV_R}{dt}.$$ (2.30)

Combination of eqs. (2.29) and (2.30), together with the application of eq. (2.7) results in

$$RC\frac{dV_R}{dt} = -V_R.$$ (2.31a)

By introducing a *new variable* t/RC, eq. (2.31a) becomes

$$\frac{dV_R}{d(t/RC)} = -V_R,$$ (2.31b)

since RC is constant.

Thus, according to the preceding section,

$$V_R = Ae^{-t/RC},$$ (2.32)

where A remains to be determined. We already assumed that $(V_C)_{t \leqslant 0} = 0$. Also, according to eq. (2.7), a sudden change in the voltage across the capacitor would imply an infinite current I which is not available from a finite V_{in} through a non-zero R. Thus immediately after $t = 0$ the voltage across the capacitor is still zero: $(V_C)_{t=0^+} = 0$, hence also $(V_R)_{t=0^+} = (V_{in})_{t=0^+} - (V_C)_{t=0^+} = V_0$. However, by use of eq. (2.32), $(V_R)_{t=0^+} = A$, hence $A = V_0$ and for times $t > 0$ we have $V_R = V_0 e^{-t/RC}$. Therefore, since V_{in}, V_R, and V_C are all zero for $t < 0$,

$$V_R = V_0 u(t)e^{-t/RC}.$$ (2.33)

Also, by use of eq. (2.28), we get

$$V_C = V_0 u(t)(1 - e^{-t/RC}).$$ (2.34)

Equations (2.33) and (2.34), shown in the lower graph of Figure 2.17, provide the transients of V_R and V_C in the circuit of Figure 2.16a for a step voltage input $V_{in} = V_0 u(t)$. We can see that immediately after time $t = 0$ the voltage across the capacitor $V_C = 0$, and the full source voltage V_0 appears across the resistor. As time increases, capacitor voltage V_C tends toward V_0, and the voltage across the resistor diminishes. In the equilibrium at $t \rightarrow \infty$, $V_C = V_0$, $V_R = 0$, and $I = 0$. For times between $t = 0$ and $t = \infty$, V_R and V_C are given by eqs. (2.33) and (2.34), respectively.

Example 2.7 In the circuit of Figure 2.16a, $V_{in} = 1$ volt $\times u(t)$, and the initial voltage across the capacitor is zero. Component values are $R = 1$ kΩ and $C = 1$ μF $= 10^{-6}$ farad. What are the values of V_R and V_C at time $t = 2$ milliseconds?

We have $RC = 1$ kΩ $\times 1$ μF $= 1$ millisecond. Thus $t/RC = 2$ and $e^{-t/RC} = e^{-2} = 0.135$. Thus, by use of eq. (2.33), $(V_R)_{t=2 \text{ msec}} = 0.135$ volt. Also, by use of eq. (2.34), $(V_C)_{t=2 \text{ msec}} = 0.865$ volt.

2.6.2 Pulse Voltage Input

Now we turn to the transient response of the circuit of Figure 2.16a in the case when V_{in} is a pulse. Following the principle of *linear superposition* we decompose the pulse into two step voltages according to Figure 2.13, compute the transient response originating from each of the two step voltages, and obtain the resulting transient as the sum of the two responses.

Example 2.8 The input voltage V_{in} in the circuit of Figure 2.16a is a pulse that starts at $t = 0$, ends at $t = 2$ milliseconds, and has a height of 1 volt as illustrated by the solid line in Figure 2.18. Component values are $R = 1$ kΩ and $C = 1$ μF, thus $RC = 1$ millisecond.

The pulse can be decomposed into two parts as $V_{in} = V_{in_1} + V_{in_2}$, where $V_{in_1} = V_0 u(t)$ and $V_{in_2} = -V_0 u(t - W)$ with $V_0 = 1$ volt and $W = 2$ milliseconds. We denote by V_{C_1} the voltage across the capacitor that originates from V_{in_1} and by V_{C_2} the voltage across the capacitor that originates from V_{in_2}. By use of eq. (2.34) we get

$$V_{C_1} = V_0 u(t) [1 - e^{-t/RC}]$$

and

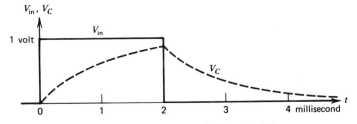

Figure 2.18 Transients in Example 2.8.

$$V_{C_2} = -V_0 u(t - W) [1 - e^{-(t-W)/RC}].$$

Hence,

$$V_C = V_{C_1} + V_{C_2} = V_0 u(t) [1 - e^{-t/RC}] - V_0 u(t - W) [1 - e^{-(t-W)/RC}]$$

where $V_0 = 1$ volt, $W = 2$ milliseconds, and $RC = 1$ millisecond. Thus for times between $t = 0$ and $t = W$,

$$V_C = V_0 (1 - e^{-t/RC})$$

and for times $t \geq W$, V_C becomes

$$V_C = V_0 (1 - e^{-W/RC}) e^{-(t-W)/RC}.$$

The resulting V_C is shown by broken line in Figure 2.18.

In general, if the voltage across the capacitor is given at a time $t = t_1$ and if v_{in} is a constant V_1 for times $t > t_1$, then it can be shown that for $t \geq t_1$ voltage V_C can be written as

$$(V_C)_{t \geq t_1} = (V_C)_{t=t_1} + [V_1 - (V_C)_{t=t_1}] [1 - e^{-(t-t_1)/RC}]$$

$$= V_1 - [V_1 - (V_C)_{t=t_1}] e^{-(t-t_1)/RC}. \qquad (2.35)$$

Thus, according to eq. (2.35), $V_C = (V_C)_{t=t_1}$ at time $t = t_1$, and $V_C = V_1$ for $t \to \infty$, as expected.

Example 2.9   In Example 2.8 and in Figure 2.18 the voltage across the capacitor at time $t_1 = 2$ milliseconds is $(V_C)_{t=t_1} = 0.865$ volt; also $V_{in} = 0$ for time $t > t_1$. Thus, by use of eq. (2.35) with $V_1 = 0$, we can write V_C for $t \geq t_1$ as

$$(V_C)_{t \geq t_1} = (V_C)_{t=t_1} e^{-(t-t_1)/RC},$$

in agreement with Figure 2.18.

2.6.3 Ramp Voltage Input

The solid line in Figure 2.19 shows a *ramp voltage*. It can be shown[†] that when a ramp voltage of $V_{in} = u(t) V_0 t/RC$ is applied as an input voltage in the circuit of Figure 2.16a, then the voltage across the capacitor can be written as

$$V_C = u(t) V_0 \left(\frac{t}{RC} - 1 + e^{-t/RC} \right) \qquad (2.36)$$

as shown by broken line in Figure 2.19.

[†]By taking the integral of eq. (2.34).

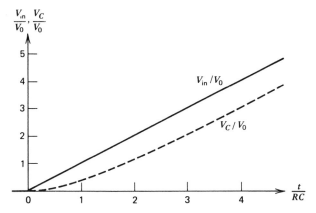

Figure 2.19 A ramp voltage V_{in} (solid line) applied as input in the R-C circuit of Figure 2.16*a* and the resulting V_C (broken line).

Example 2.10 The solid line in Figure 2.20 shows a pulse V_{in} with finite rise and fall times. We use this pulse as input in the circuit of Figure 2.16*a*. For $0 \leqslant t/RC \leqslant 2$, $V_0 = 0.5$ volt and, by applying eq. (2.36),

$$V_C = 0.5 \text{ volt} \left(\frac{t}{RC} - 1 + e^{-t/RC} \right)$$

as shown by broken line in Figure 2.20. At $t/RC = 2$

$$V_C(t/RC = 2) = 0.5 \text{ volt} (2 - 1 + e^{-2}) = 0.57 \text{ volt.}$$

Thus for $2 \leqslant t/RC \leqslant 6$, according to eq. (2.35),

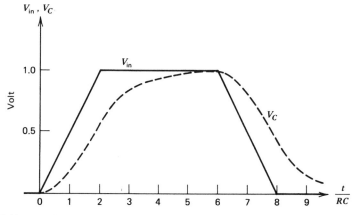

Figure 2.20 A pulse V_{in} with finite rise and fall times (solid line) applied as input in the R-C circuit of Figure 2.16*a* and the resulting V_C (broken line).

$$V_C = 1 \text{ volt} - (1 \text{ volt} - 0.57 \text{ volt}) \, e^{-(t/RC-2)}$$

$$= 1 \text{ volt} - 0.43 \text{ volt } e^{-(t/RC-2)}.$$

At $t/RC = 6$, $V_C = 1 \text{ volt} - 0.43 \text{ volt } e^{-(6-2)} = 1 \text{ volt} - 0.0077 \text{ volt} \approx 1 \text{ volt}$. By using the result for $0 \leqslant t/RC \leqslant 2$, we get for V_C during $6 \leqslant t/RC \leqslant 8$

$$V_C = 1 \text{ volt} - 0.5 \text{ volt} \left[\left(\frac{t}{RC} - 6 \right) - 1 + e^{-(t/RC-6)} \right].$$

Finally, for $t/RC \geqslant 8$

$$V_C = 0.43 \text{ volt } e^{-(t/RC-8)}.$$

2.6.4 Sinewave Input

Now we consider the circuit of Figure 2.16a with a *sinewave* input of

$$V_{in} = V_0 \sin (2\pi ft + \phi_0). \tag{2.37}$$

In eq. (2.37) and in what follows f is the *frequency* of the sinewave (measured in Hertz, Hz, which is the same as cycles per second, cps: 1 Hz = 1 cps). The *initial phase* is ϕ_0; it can be shown that we can arbitrarily equate $\phi_0 = 0$ without any loss of generality. Thus, eq. (2.37) becomes

$$V_{in} = V_0 \sin 2\pi ft. \tag{2.38}$$

We do not derive voltage V_C in the circuit of Figure 2.16a for the V_{in} of eq. (2.38), but only give the result:

$$V_C = \frac{V_0}{\sqrt{1 + (2\pi RCf)^2}} \sin (2\pi ft - \arctan 2\pi RCf). \tag{2.39}$$

Next we introduce *corner frequency* f_0 defined as

$$f_0 = \frac{1}{2\pi RC}. \tag{2.40}$$

It is also customary to separate the *magnitude* $|V_C|$ and the *phase* $\underline{/V_C}$ of voltage V_C, and write V_C as

$$V_C = |V_C| \sin (2\pi ft + \underline{/V_C}) \tag{2.41}$$

where

$$|V_C| = \frac{|V_0|}{\sqrt{1 + (2\pi RCf)^2}} \tag{2.42a}$$

and

$$\underline{/V_C} = -\arctan 2\pi RCf. \tag{2.42b}$$

With the corner frequency of eq. (2.40), eqs. (2.42a) and (2.42b) become

$$|V_C| = \frac{|V_0|}{\sqrt{1 + (f/f_0)^2}} \tag{2.43}$$

and

$$\underline{/V_C} = -\arctan f/f_0. \tag{2.44}$$

Equations (2.43) and (2.44) are illustrated in the *frequency response* graphs of Figure 2.21 using logarithmic scales for $|V_C/V_0|$ and f/f_0. Several special cases can be identified. (1): $f/f_0 = 1$, resulting in $|V_C/V_0| = 1/\sqrt{2}$ and $\underline{/V_C} = -45°$; (2): $f/f_0 \gg 1$, leading to $|V_C/V_0| \approx f_0/f$ and $\underline{/V_C} \approx -90°$; (3): $f/f_0 \ll 1$, in which case $|V_C/V_0| \approx 1 - \frac{1}{2}(f/f_0)^2$ and $\underline{/V_C} \approx -f/f_0$ (in radians).

In the case when $f \ll f_0$, the approximations for the magnitude and phase can be also written as

$$|V_C| \approx |V_0|(1 - \pi t_R^2 f^2) \tag{2.45a}$$

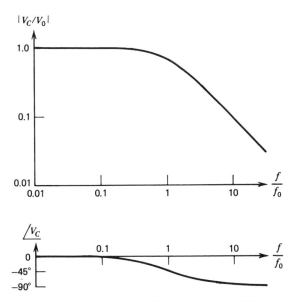

Figure 2.21 Magnitude and phase of voltage V_C in the circuit of Figure 2.16a with a sine-wave input.

and

$$\underline{/V_C} \approx -2\pi f t_D, \tag{2.45b}$$

where t_D and t_R are, respectively, the *Elmore delay* and the *Elmore risetime* given by

$$t_D = RC \tag{2.45c}$$

and

$$t_R = \sqrt{2\pi}\, RC. \tag{2.45d}$$

Example 2.11   The input voltage in the circuit of Figure 2.16a is given by eq. (2.38) with V_0 = 5 volts and f = 60 Hz; circuit values are R = 100Ω and C = 1 μF. Thus, f_0 = $1/(2\pi RC)$ = 1592 Hz $>> f$ = 60 Hz, hence eqs. (2.45) are applicable. We get $t_D = RC$ = 100 Ω × 1 μF = 0.1 msec, $t_R = \sqrt{2\pi}\, RC$ = 0.25 msec, and, by use of eqs. (2.41) and (2.45), $V_C \approx 0.9993 \times$ 5 volt × sin $[2\pi$ 60 Hz × $(t - 0.1$ msec$)]$.

Next we look at voltage V_R across resistor R in the circuit of Figure 2.16a with V_{in} given by eq. (2.38). Since $V_{in} = V_R + V_C$, V_R can be obtained as the difference of eq. (2.38) and eq. (2.39). After some algebraic manipulations we get

$$\frac{V_R}{V_0} = \frac{2\pi RCf}{\sqrt{1 + (2\pi RCf)^2}}\, \sin\,(2\pi ft + \operatorname{arccot} 2\pi RCf). \tag{2.46}$$

By introducing again corner frequency f_0 as

$$f_0 = \frac{1}{2\pi RC}, \tag{2.47}$$

the magnitude and the phase of V_R become

$$|V_R| = \frac{|V_0|}{\sqrt{1 + (f_0/f)^2}}, \tag{2.48}$$

$$\underline{/V_R} = \arctan f_0/f. \tag{2.49}$$

Equations (2.48) and (2.49) are illustrated in Figure 2.22 using logarithmic scales for $|V_R/V_0|$ and f/f_0. Several special cases can be identified. (1): $f/f_0 = 1$, resulting in $|V_R/V_0| = 1/\sqrt{2}$ and $\underline{/V_R}$ = 45°; (2): $f/f_0 << 1$, leading to $|V_R/V_0| \approx f/f_0$ and $\underline{/V_R} \approx$ 90°; (3): $f/f_0 >> 1$, in which case $|V_R/V_0| \approx 1 - \frac{1}{2}(f_0/f)^2$ and $\underline{/V_R} \approx f_0/f$ (in radians).

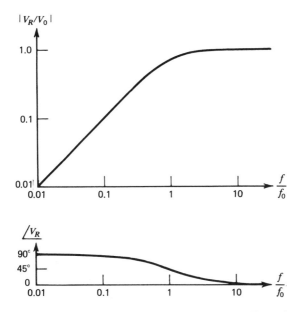

Figure 2.22 Magnitude and phase of voltage V_R in the circuit of Figure 2.16a with a sine-wave input.

2.7 R-L CIRCUITS

Figure 2.23 shows two forms of an R-L circuit that are equivalent if $I_{in} = V_{in}/R$. Because the equations describing the behavior of inductors are similar to those for capacitors, derivations of transients in R-L circuit are not performed here. However, we give here the transient of V_R in the circuit of Figure 2.23a for the case when $V_{in} = V_0 u(t)$:

$$V_R = V_0 u(t)\,(1 - e^{-tR/L}).\qquad(2.50)$$

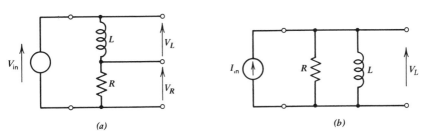

Figure 2.23 Two forms of an R-L circuit that are equivalent if $I_{in} = V_{in}/R$.

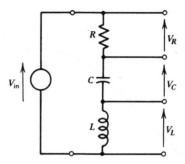

Figure 2.24   A series R-L-C circuit.

2.8 SERIES R-L-C CIRCUITS

Figure 2.24 shows a series R-L-C circuit driven by a voltage source. The response of the circuit is described for a step voltage input, for a step voltage input with a finite risetime, and for a sinewave.

2.8.1 Step Voltage Input

For a step voltage input

$$V_{in} = V_0 u(t), \tag{2.51}$$

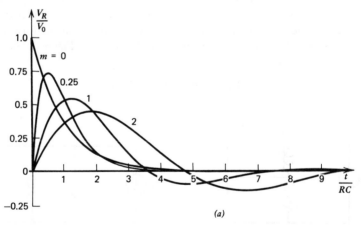

Figure 2.25   Voltages in the circuit of Figure 2.24 for a step voltage input of $V_{in} = V_0 u(t)$ and for various values of $m = L/R^2C$. (a) Voltage across the resistor; (b) voltage across the inductor; (c) voltage across the capacitor.

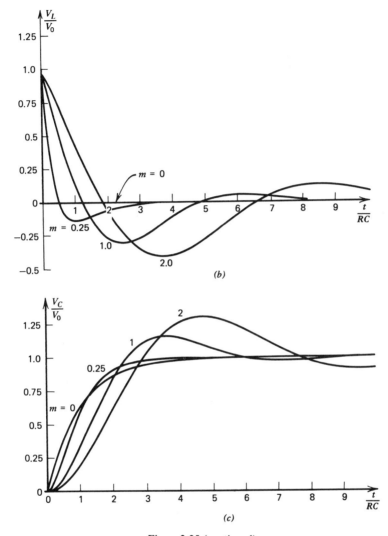

Figure 2.25 (continued)

the response of the circuit can be shown to be as depicted in Figure 2.25, where parameter m is defined as

$$m = \frac{L}{R^2 C}.$$
(2.52)

Four cases can be distinguished: $m = 0$, $0 < m < 0.25$, $m = 0.25$, and $m > 0.25$. In all four cases, $V_C/V_0 + V_R/V_0 + V_L/V_0 = 1$ for $t > 0$.

Case of $m = 0$. In this case the circuit reduces to the circuit of Figure 2.16a: V_C is given by eq. (2.34), V_R by eq. (2.33), and $V_L = 0$.

Case of $0 < m < 0.25$. It can be shown that in this *overdamped* case, V_R and V_L are given by

$$\frac{V_R}{V_0} = u(t)\frac{e^{-(1-\sqrt{1-4m})t/2mRC} - e^{-(1+\sqrt{1-4m})t/2mRC}}{\sqrt{1-4m}} \qquad (2.53)$$

$$\frac{V_L}{V_0} = 0.5\,u(t)$$

$$\times \frac{[(1+\sqrt{1-4m})\,e^{-(1+\sqrt{1-4m})t/2mRC} - (1-\sqrt{1-4m})\,e^{-(1-\sqrt{1-4m})t/2mRC}]}{\sqrt{1-4m}}.$$

$$(2.54)$$

Case of $m = 0.25$. This is the case of *critical damping* and

$$\frac{V_R}{V_0} = 4u(t)\frac{t}{RC}e^{-2t/RC}, \qquad (2.55)$$

$$\frac{V_L}{V_0} = u(t)\left(1 - \frac{2t}{RC}\right)e^{-2t/RC}. \qquad (2.56)$$

Case of $m > 0.25$. In this *underdamped* (or *oscillatory*) case,

$$\frac{V_R}{V_0} = u(t)\frac{e^{-t/2mRC}\sin(\sqrt{m-0.25}\,t/mRC)}{\sqrt{m-0.25}}, \qquad (2.57)$$

$$\frac{V_L}{V_0} = u(t)e^{-t/2mRC}\left[\cos(\sqrt{m-0.25}\,t/mRC) - \frac{\sin(\sqrt{m-0.25}\,t/mRC)}{2\sqrt{m-0.25}}\right].$$

$$(2.58)$$

We can see that the transient of V_C is monotonic when $m \leqslant 0.25$. However, even here V_L has an initial value of V_0 and a finite undershoot for any $m \neq 0$ as summarized in Figure 2.26, which may be detrimental in some applications.

Example 2.12 Figure 2.27 shows two instruments with their grounds interconnected by a copper strip that has a negligible resistance, but has an inductance of $L = 0.18\,\mu H$. A signal is transmitted between the two instruments by way of a coaxial cable that can be represented by a capacitance of $C = 72$ pF. Thus, $RC = 7.2$ nsec, $m = L/R^2C = 0.25$, and the voltage drop V_L on the ground return is given by the $m = 0.25$ curve in Figure

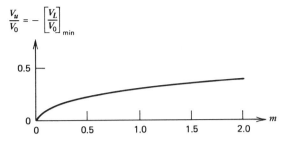

Figure 2.26   Undershoot V_u of inductor voltage V_L in Figure 2.25b as function of m = L/R^2C.

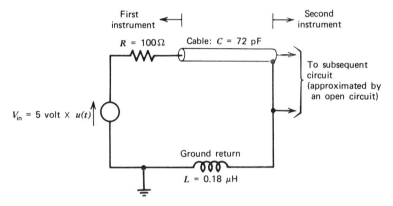

Figure 2.27   Two instruments interconnected by a ground return with inductance L.

2.25b. We can see that the full 5 volt voltage swing appears across the ground return at time $t = 0$, and it becomes zero at time $t = 0.5\,RC = 3.6$ nsec; also, the subsequent undershoot has a maximum magnitude of 0.14×5 volt = 0.7 volt at time $t = RC = 7.2$ nsec. The voltage V_L on the ground return also appears at the input of other circuits not shown in Figure 2.27 that assume a ground connection between the two instruments with $V_L \approx 0$. These circuits may be sensitive enough to respond to voltage V_L, resulting in erroneous operation.

2.8.2   Step Voltage Input with Finite Risetime

A step voltage with a finite risetime is shown in Figure 2.28. It can be described as

$$V_{in} = V_0\,u(t)\,\frac{t}{t_r} - V_0\,u(t - t_r)\,\frac{t - t_r}{t_r}, \qquad (2.59)$$

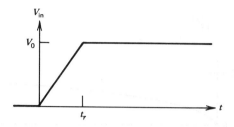

Figure 2.28   Step voltage with height V_0 and risetime t_r.

where t_r is *not* identical with Elmore risetime t_R, or with a 10 to 90% risetime.

The resulting voltages in the circuit of Figure 2.24 are too long to be printed here when V_{in} of eq. (2.59) is applied as input. However, the resulting V_L is illustrated in Figure 2.29 for $t_r/RC = 0.5$ and 2. We can see by comparing these two and also V_L/V_0 of Figure 2.25b, that for a given m an increasing t_r reduces the magnitudes of both the maximum and the minimum of V_L. These magnitudes are summarized in Figures 2.30 and 2.31 as functions of t_r/RC for several values of m.

Example 2.13   The voltage source of Example 2.12, shown in Figure 2.27, is replaced by one that has a finite risetime of 3.6 nsec; that is, it is described by eq. (2.59) with $V_0 = 5$ volts (same as before) and $t_r = 3.6$ nsec. Thus $t_r/RC = 0.5$; also we still have $m = 0.25$. From Figure 2.30 we get a $V_{L_{max}} = 1.9$ volts as compared with the 5 volts in Example 2.12. Also, from Figure 2.31 we get a $- V_{L_{min}} = 0.65$ volt as compared with the 0.7 volt in Example 2.12. Thus the finite risetime reduces the voltage drop on the ground return to a more acceptable level.

2.8.3 Sinewave Input

We look only at the voltage across the capacitor in Figure 2.24 when the input is a sinewave. When

$$V_{in} = V_0 \sin 2\pi ft, \tag{2.60}$$

the resulting magnitude of V_C is

$$|V_C| = \frac{|V_0|}{\sqrt{[1 - m(f/f_0)^2]^2 + (f/f_0)^2}}, \tag{2.61a}$$

and the resulting phase of V_C is

$$\underline{/V_C} = - \arctan\frac{f/f_0}{1 - m(f/f_0)^2}, \tag{2.61b}$$

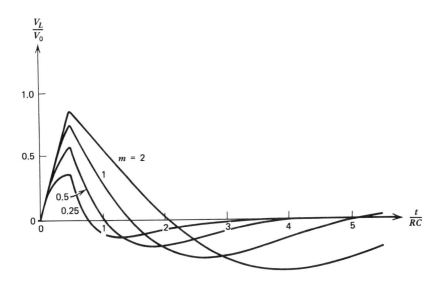

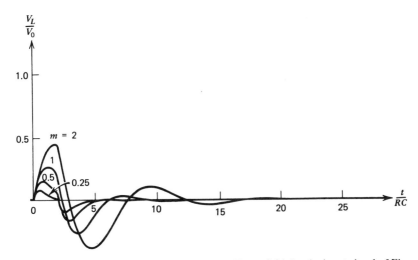

Figure 2.29 Inductor voltage V_L in the circuit of Figure 2.24 for the input signal of Figure 2.28 with various values of $m = L/R^2C$. (a) $t_r = 0.5\ RC$; (b) $t_r = 2\ RC$. Note different time scales.

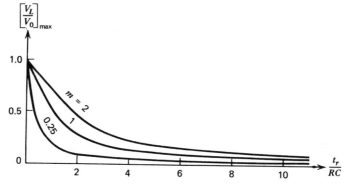

Figure 2.30 Maximum values of inductor voltage V_L in Figure 2.29 as function of t_r/RC for various values of $m = L/R^2C$.

where corner frequency f_0 is given by

$$f_0 = \frac{1}{2\pi RC}. \tag{2.61c}$$

Equation (2.61) is illustrated in Figure 2.32 for various values of m. The magnitude is monotonic when $m \leqslant 0.5$, and it has a resonance peak for $m > 0.5$. For frequencies of $f \ll f_0$ the magnitude and the phase of V_C can be approximated as

$$|V_C| \approx |V_0| (1 - \pi t_R^2 f^2) \tag{2.62a}$$

and

$$\underline{/V_C} \approx -2\pi f t_D, \tag{2.62b}$$

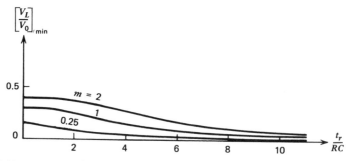

Figure 2.31 Minimum values of inductor voltage V_L in Figure 2.29 as function of t_r/RC for various values of $m = L/R^2C$.

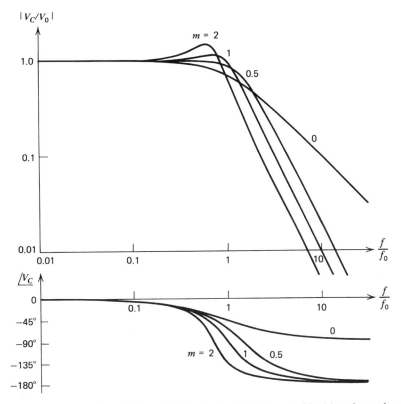

Figure 2.32 Magnitude and phase of V_C in the circuit of Figure 2.24 with various values of $m = L/R^2C$ for a sinewave input of $V_{in} = V_0 \sin 2\pi ft$.

where t_D and t_R are, respectively, the Elmore delay and the Elmore risetime that are now given by[†]

$$t_D = RC \qquad (2.62c)$$

and

$$t_R = \sqrt{2\pi} \sqrt{1 - 2m} \, RC. \qquad (2.62d)$$

Thus when $m < 0.5$, $t_R^2 > 0$ and according to eq. (2.62a) the initial slope of $|V_C|$ versus f^2 is negative. When $m > 0.5$, $t_R^2 < 0$, t_R does not exist, and the initial slope is positive leading up to the resonance peak (see Figure 2.32). In

[†]In general, if the transfer function in the complex frequency domain is given as constant $\times$ $(1 + a_1s + a_2s^2 + \cdots)/(1 + b_1s + b_2s^2 + \cdots)$ where $s = j2\pi f$, then $t_D = b_1 - a_1$ and $t_R^2 = 2\pi(b_1^2 - a_1^2 - 2b_2 + 2a_2)$.

the special case of $m = 0.5$, $t_R = 0$ and the frequency-response magnitude as function of frequency is *maximally flat*.

2.8.4 Delay, Risetime, and Overshoot

The transient of V_C in Figure 2.24 for a step input V_{in} was shown in Figure 2.25c for various values of m. While, of course, such graphs completely characterize the transients, parameters that provide condensed information are often adequate. One such parameter is the delay of the 50% point, t_{50}, which is the time it takes the transient to reach 50% of its final value. Another parameter is the 10 to 90% risetime, t_{10-90}, which is the time the transient spends between 10 and 90% of its final value. A third one is the fractional overshoot ϵ, which is the magnitude of the overshoot divided by the final value of the transient.

When $m = 0$, it can be shown that t_{50} and t_{10-90} are given as $t_{50} = RC$ $\ln 2 \approx 0.7\ RC$ and $t_{10-90} = RC \ln 9 \approx 2.2\ RC$; also, $\epsilon = 0$ when $m \leqslant 0.25$. In general, however, these have to be read from the graphs of the transients. The t_{50}, t_{10-90}, and ϵ thus obtained are shown in Figure 2.33 for $0 \leqslant m \leqslant 2$; t_D and t_R are also shown (for $m > 0.5$, $t_R^2 < 0$ hence t_R does not exist).

Note that while t_{50}, t_{10-90}, and ϵ have to be read from the transients (or have to be measured), Elmore delay t_D and Elmore risetime t_R (which are low frequency characteristics) can be readily computed from eqs. (2.62c) and (2.62d). It is often desirable to have some relationships between the transient characteristics t_{50} and t_{10-90} and the low frequency characteristics t_D and t_R, no matter how crude and limited these may be. It can be shown that when the transient response is free of overshoot *and* ringing, that is, when the transient is *monotonic*,

$$t_{50} \approx t_D \qquad\qquad (2.63a)$$

and

$$t_{10-90} \approx t_R \qquad\qquad (2.63b)$$

provide crude but reasonable approximations.

Note that the lack of overshoot, that is, an $\epsilon = 0$ criterion is insufficient for monotonicity, since there may be ringing in the transient making it nonmonotonic even when the overshoot $\epsilon = 0$, as shown in the graph below. With the

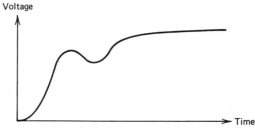

A nonmonotonic voltage that has an overshoot of $\epsilon = 0$.

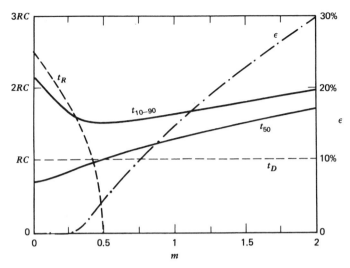

Figure 2.33 Elmore delay t_D, Elmore risetime t_R, delay of the 50% point t_{50}, 10 to 90% risetime t_{10-90}, and overshoot ϵ in the circuit of Figure 2.24 for various values of $m = L/R^2C$.

requirement of $\epsilon = 0$ hence $m \leqslant 0.25$, the worst disagreements of eqs. (2.63) in Figure 2.33 are at $m = 0$, where $t_{50} \approx 0.7\,RC$ and $t_D = RC$, and also $t_{10-90} \approx 2.2\,RC$ and $t_R \approx 2.5\,RC$. We shall see later that in reality this is just about as bad as the approximations of eqs. (2.63) ever get, provided that the transient is indeed monotonic.

2.9 PARALLEL-SERIES R-L-C CIRCUITS

Figure 2.34 shows one of the several possible parallel-series R-L-C circuits. The response of the circuit is described here for step current input and for sinewave input.

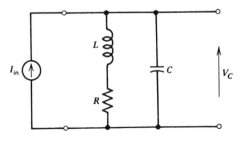

Figure 2.34 A parallel-series R-L-C circuit.

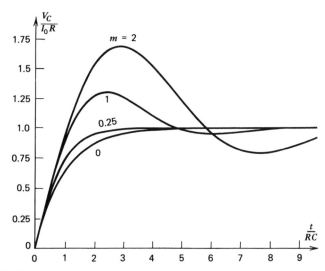

Figure 2.35 Voltage V_C across the capacitor in the circuit of Figure 2.34 with a current input of $I_{in} = I_0 u(t)$.

2.9.1 Step Current Input

When the input is a step current

$$I_{in} = I_0 u(t), \tag{2.64}$$

voltage V_C across the capacitor is given by Figure 2.35 where, as before,

$$m = \frac{L}{R^2 C}. \tag{2.65}$$

The transients of Figure 2.35 are somewhat similar, but not identical, to those of the series circuit illustrated in Figure 2.25c. However, the transients in Figure 2.35 reduce to those in Figure 2.17 when $m = 0$, as expected.

2.9.2 Sinewave Input

When the input current in the circuit of Figure 2.34 is a sinewave

$$I_{in} = I_0 \sin 2\pi ft, \tag{2.66}$$

the resulting magnitude and phase of V_C can be written as

$$|V_C| = |I_0| R \sqrt{\frac{1 + (mf/f_0)^2}{[1 - m(f/f_0)^2]^2 + (f/f_0)^2}}, \tag{2.67}$$

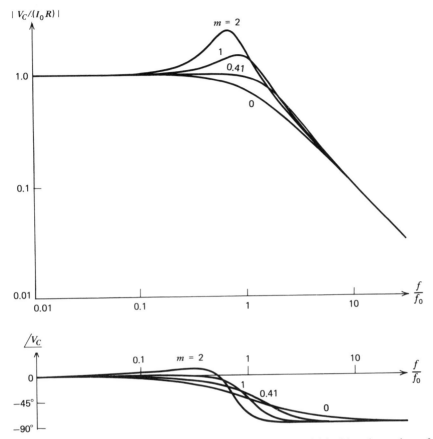

Figure 2.36 Magnitude and phase of V_C in the circuit of Figure 2.34 with various values of $m = L/R^2C$ for a sinewave input of $I_{in} = I_0 \sin 2\pi ft$.

$$\underline{/V_C} = -\arctan \{[1 - m + (mf/f_0)^2] f/f_0\}, \tag{2.68}$$

where, as before,

$$f_0 = \frac{1}{2\pi RC}. \tag{2.69}$$

Equations (2.67) and (2.68) are illustrated in Figure 2.36. We can see that the magnitude versus frequency plot is monotonic when $m \leqslant \sqrt{2} - 1 \approx 0.41$ and it has a resonance peak for $m > 0.41$. For frequencies of $f \ll f_0$, the magnitude and the phase of V_C can be approximated as

$$|V_C| \approx |V_0| (1 - \pi t_R^2 f^2) \qquad (2.70a)$$

and

$$\underline{/V_C} \approx -2\pi f t_D, \qquad (2.70b)$$

where t_D and t_R are, respectively, the Elmore delay and the Elmore risetime, given by

$$t_D = (1 - m)RC \qquad (2.70c)$$

and

$$t_R = \sqrt{2\pi} \sqrt{1 - m^2 - 2m} RC. \qquad (2.70d)$$

Thus when $m \leqslant 0.41$, $t_R^2 > 0$, and according to eq. (2.70a) the initial slope of $|V_C|$ versus f^2 is negative. When $m > 0.41$, $t_R^2 < 0$, t_R does not exist, and the initial slope is positive leading up to the resonance peak in Figure 2.36. In the special case of $m = \sqrt{2} - 1 = 0.41$, $t_R = 0$, and the magnitude as function of frequency is maximally flat.

2.9.3 Delay, Risetime, and Overshoot

The t_{50}, t_{10-90}, ϵ, t_D, and t_R for the V_C of Figure 2.34 are shown in Figure 2.37. As before, there is a crude agreement between t_{50} and t_D as well as between t_{10-90} and t_R when $m \leqslant 0.25$, that is, when the transient is free from

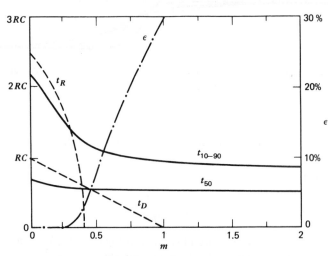

Figure 2.37 Elmore delay t_D, Elmore risetime t_R, delay of the 50% point t_{50}, 10 to 90% risetime t_{10-90}, and overshoot ϵ in the circuit of Figure 2.34 for various values of $m = L/R^2C$.

overshoot and ringing. The use of Figure 2.37 is illustrated in the example that follows.

Example 2.14 In the parallel-series R-L-C circuit of Figure 2.34, $R = 100$ Ω, $C = 36$ pF, and $L = 0.18$ μH, thus $RC = 3.6$ nsec and $m = L/R^2C = 0.5$. Initially all voltages and currents in the circuit are zero. I_{in} is a step current $I_{in} = 10$ mA $\times$ $u(t)$, thus the final voltage across the capacitor is 10 mA $\times$ 100 Ω = 1 volt. For $m = 0.5$, from Figure 2.37, $t_{50} = 0.55\ RC$, $t_{10-90} = 1.1\ RC$, and $\epsilon = 7.5\%$. Hence the voltage across the capacitor reaches 0.5 volt at time $0.55\ RC \approx 2$ nsec, spends between 0.1 and 0.9 volt a time of $1.1\ RC \approx 4$ nsec, and has an overshoot of 1 volt $\times$ 7.5% = 75 mV.

2.10 PULSE TRANSFORMERS

The subject of pulse transformers is an extensive and difficult one. In this section we merely present two simple equivalent circuits and show the resulting responses for step voltage and sinewave inputs.

Figure 2.38 shows the schematic diagram of a pulse transformer with two windings that are usually wound on a toroidal magnetic core. One winding consists of N_1 turns, the other winding of N_2 turns. An *ideal transformer* does not dissipate any power, hence, in Figure 2.38

$$V_1 I_1 = V_2 I_2, \tag{2.71a}$$

$$\frac{V_2}{V_1} = \frac{N_2}{N_1}, \tag{2.71b}$$

and

$$\frac{I_2}{I_1} = \frac{N_1}{N_2}; \tag{2.71c}$$

further, in an ideal transformer eqs. (2.71) hold for all times and also for all frequencies *including dc.*

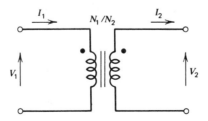

Figure 2.38 Schematic diagram of a pulse transformer.

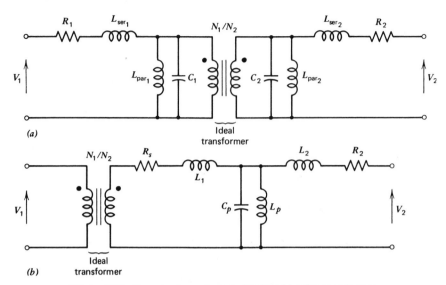

(a)

(b)

Figure 2.39 Two simple equivalent circuits of a pulse transformer.

Two equivalent circuits of a *real transformer* are shown in Figure 2.39. In Figure 2.39a, resistances R_1 and R_2 represent the series resistances of the windings, L_{par_1} and L_{par_2} their inductances, C_1 and C_2 their capacitances, and L_{ser_1} and L_{ser_2} their leakage inductances. In a well-designed pulse transformer, $L_{\mathrm{ser}_1} \ll L_{\mathrm{par}_1}$ and $L_{\mathrm{ser}_2} \ll L_{\mathrm{par}_2}$. The components of Figure 2.39b are related to those of Figure 2.39a by $R_s = R_1 \, N_2^2/N_1^2$, $L_1 = L_{\mathrm{ser}_1} \, N_2^2/N_1^2$, $C_p = C_2 + C_1 \, N_1^2/N_2^2$, $1/L_p = 1/L_{\mathrm{par}_2} + N_1^2/(N_2^2 \, L_{\mathrm{par}_1})$, and $L_2 = L_{\mathrm{ser}_2}$.

In the simplest case, a pulse transformer is driven by a voltage source in series with a source resistance R_G and is loaded by a load resistance R_L. We show the response of this circuit under the simplifying assumptions that in Figure 2.39b $R_2 + R_L = R_s + R_G N_2^2/N_1^2 = R$, $L_1 = L_2 = L$, and that we omit L_p and thereby restrict our attention to times that are much shorter than L_p/R and to frequen-

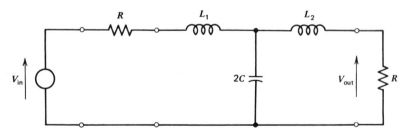

Figure 2.40 Simplified pulse transformer circuit for the computation of the transient and sinewave responses.

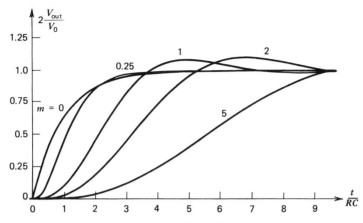

Figure 2.41 Transient response of the circuit of Figure 2.40 for various values of $m = L/R^2C$.

cies that are much higher than $R/(2\pi L_p)$. With these assumptions and with $C = C_p/2$, we arrive at the circuit shown in Figure 2.40.

2.10.1 Step Voltage Input

When V_{in} is a step voltage

$$V_{in} = V_0 u(t), \tag{2.72}$$

the response of the circuit of Figure 2.40 is as illustrated in Figure 2.41, where

$$m = \frac{L}{R^2 C}. \tag{2.73}$$

We can see that the transients are monotonic when $m \leqslant 0.25$, and that they exhibit overshoot and ringing when $m > 0.25$.

2.10.2 Sinewave Input

When the input is a sinewave

$$V_{in} = V_0 \sin 2\pi ft, \tag{2.74}$$

the resulting magnitude and phase of V_{out} can be written as

$$|V_{out}| = \frac{0.5 |V_0|}{\sqrt{[1 - 2m(f/f_0)^2]^2 + (f/f_0)^2 [1 + m - (mf/f_0)^2]^2}} \tag{2.75}$$

and

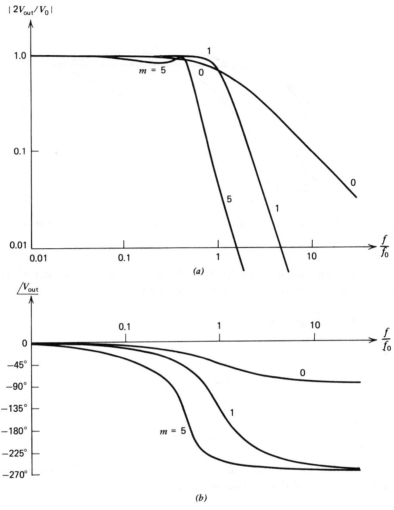

Figure 2.42 Magnitude (*a*) and phase (*b*) of V_{out} in the circuit of Figure 2.40 with various values of $m = L/R^2C$ for a sinewave input of $V_{in} = V_0 \sin 2\pi ft$.

$$\underline{/V_{out}} = -\arctan \frac{(f/f_0)\ [1 + m - (mf/f_0)^2]}{1 - 2m(f/f_0)^2}. \qquad (2.76)$$

Equations (2.75) and (2.76) are illustrated in Figure 2.42. For frequencies that are much lower than f_0, but that are still much higher than $R/(2\pi L_p)$, the magnitude and the phase of V_{out} can be approximated as

$$|V_{\text{out}}| \approx \frac{|V_0|}{2} (1 - \pi t_R^2 f^2) \qquad (2.77a)$$

and

$$\underline{/V_{\text{out}}} \approx -2\pi f t_D, \qquad (2.77b)$$

where t_D is the Elmore delay given by

$$t_D = (1 + m)RC \qquad (2.77c)$$

and t_R is the Elmore risetime given by

$$t_R = \sqrt{2\pi}|1 - m|RC. \qquad (2.77d)$$

We can see that t_R^2 is never negative, hence there is no resonance peak for any m in the upper (magnitude) graph of Figure 2.42. For $m > 1$, however, the magnitude is not monotonic, and the $m = 1$ case represents the maximally flat frequency-response magnitude.

2.10.3 Delay, Risetime, and Overshoot

Values of t_{50}, t_{10-90}, ϵ, t_D, and t_R are summarized in Figure 2.43. As before, there is a crude agreement between t_{50} and t_D as well as between t_{10-90} and t_R when $m \leqslant 0.25$, that is, when the transient is free from overshoot and ringing.

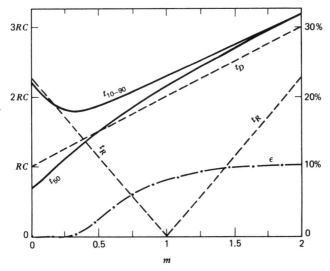

Figure 2.43 Elmore delay t_D, Elmore risetime t_R, delay of the 50% point t_{50}, 10 to 90% risetime t_{10-90}, and overshoot ϵ in the circuit of Figure 2.40 for various values of $m = L/R^2C$.

Also note that t_{10-90} has a minimum at $m \approx 0.4$ and that t_{50} and ϵ also degrade for larger values of m; hence, in a well-designed pulse transformer $m \leqslant 0.4$.

Example 2.15   A pulse transformer is described by Figure 2.39a with $N_2/N_1 = 2, R_1 = 0.1 \ \Omega, R_2 = 0.4 \ \Omega, L_{par_1} = 10 \ \mu H, L_{par_2} = 40 \ \mu H, L_{ser_1} = 0.05 \ \mu H, L_{ser_2} = 0.2 \ \mu H, C_1 = 50$ pF, and $C_2 = 12.5$ pF. The circuit is driven at its left terminals by a voltage source in series with a source resistance of 50 Ω, and loaded on its right terminals by a load resistance of 200 Ω. Thus $R_s = 0.4 \ \Omega, L_1 = 0.2 \ \mu H, C_p = 25$ pF, $L_p = 20 \ \mu H$, and $L_2 = 0.2 \ \mu H$. We use the approximate Figure 2.40, where $R = 200.4 \ \Omega$, $L = 0.2 \ \mu H$, and $C = 12.5$ pF, whence $RC \approx 2.5$ nsec and $m = L/R^2 C \approx 0.4$. By use of Figure 2.43 we get $t_{50} = 1.25 \ RC \approx 3.1$ nsec, $t_{10-90} = 1.8 \ RC = 4.5$ nsec, and an overshoot of $\epsilon \approx 0.5\%$.

†2.11 CASCADED CIRCUITS

In the preceding sections, the input drive to each circuit was a fixed step function. a ramp, a step function with a finite risetime, or a sinewave. Often, however— such as in cascaded amplifier stages—these sources are proportional to output voltages of preceding circuits. When the input is a sinewave, the resulting overall frequency response can be obtained by multiplying the individual magnitudes and adding the individual phases. The determination of the transient response is more involved, and we deal only with two simple cases.

2.11.1 Two-Stage R-C Circuits

Consider the two-stage R-C circuit of Figure 2.44 composed of two cascaded R-C circuits of Figure 2.16a. The voltage input to the second stage V_G is controlled by output voltage V_1 of the first stage.

When input voltage V_{in} is a step voltage

†Optional material.

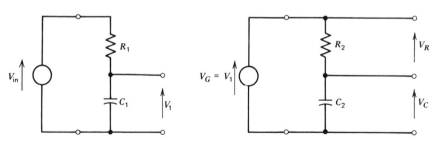

Figure 2.44   A two-stage cascaded R-C circuit.

$$V_{in} = V_0 u(t), \tag{2.78}$$

outputs V_C and V_R are governed by V_0, R_1, R_2, C_1, and C_2.

Transients of V_C. These are shown in Figure 2.45 for various values of $R_2 C_2 / R_1 C_1$. Note that $R_2 C_2 / R_1 C_1 = 0$ represents either $R_2 = 0$, or $C_2 = 0$, or both, and results in the R-C transient given by eq. (2.34) and by V_C of Figure 2.17. It can be also shown that the Elmore delay t_D of the overall transient of V_{in} to V_C in Figure 2.44 can be written as

$$t_D = t_{D_1} + t_{D_2} \tag{2.79a}$$

where t_{D_1} and t_{D_2} are the individual Elmore delays

$$t_{D_1} = R_1 C_1, \tag{2.79b}$$

$$t_{D_2} = R_2 C_2. \tag{2.79c}$$

Also, the Elmore risetime t_R of the overall transient is given by

$$t_R^2 = t_{R_1}^2 + t_{R_2}^2 \tag{2.80a}$$

or by

$$t_R = \sqrt{t_{R_1}^2 + t_{R_2}^2}, \tag{2.80b}$$

where t_{R_1} and t_{R_2} are the individual Elmore risetimes

$$t_{R_1} = \sqrt{2\pi} R_1 C_1, \tag{2.80c}$$

$$t_{R_2} = \sqrt{2\pi} R_2 C_2. \tag{2.80d}$$

The overall delay of the 50% point t_{50}, the overall 10 to 90% risetime

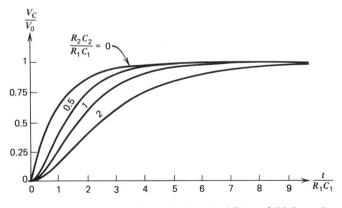

Figure 2.45 Transient response of V_C in the circuit of Figure 2.44 for various values of $R_2 C_2 / R_1 C_1$. The $R_2 C_2 / R_1 C_1 = 0$ curve represents $R_2 = 0$ and/or $C_2 = 0$.

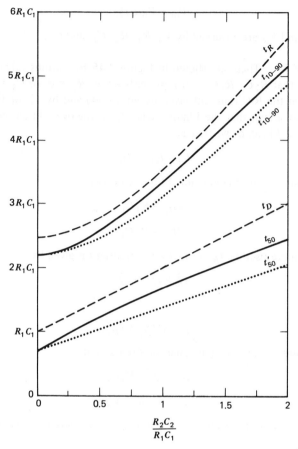

Figure 2.46 Elmore delay t_D, Elmore risetime t_R, delay of the 50% point t_{50}, 10 to 90% risetime t_{10-90}, and t_{50}' and t_{10-90}' defined by eqs. (2.81) and (2.82), all for V_C in the circuit of Figure 2.44.

t_{10-90}, the overall Elmore delay t_D, and the overall Elmore risetime t_R are shown in Figure 2.46. Also shown are t_{50}' and t_{10-90}' which are defined as

$$t_{50}' = t_{50_1} + t_{50_2}, \tag{2.81}$$

$$t_{10-90}' = \sqrt{t_{10-90_1}^2 + t_{10-90_2}^2}. \tag{2.82}$$

Delay t_{50}' approximates the overall delay t_{50} by the sum of the delay of the first stage t_{50_1} and the delay of the second stage t_{50_2}. Risetime t_{10-90}' approximates the overall risetime t_{10-90} by combining the risetime of the first stage t_{10-90_1} and the risetime of the second stage t_{10-90_2} as the square root of the sum of the squares. We can see that for the transient of V_C in Figure 2.44, reasonable approximations of t_{50} are provided by either t_D or by t'_{50}, and reasonable approximations of $t_{10\text{-}90}$ are provided by either t_R or by $t'_{10\text{-}90}$.

Example 2.16 In the circuit of Figure 2.44, $R_1 = 100\ \Omega$, $C_1 = 20$ pF, $R_2 = 100\ \Omega$, and $C_2 = 30$ pF. Thus, $R_1 C_1 = 2$ nsec, $R_2 C_2 = 3$ nsec, and $R_2 C_2 / R_1 C_1 = 1.5$. The first stage of the circuit contributes a 10 to 90% risetime of $t_{10\text{-}90_1} = 2$ nsec ln 9 = 4.4 nsec. The second stage contributes a 10 to 90% risetime of $t_{10\text{-}90_2} = 3$ nsec ln 9 = 6.6 nsec. The exact value of the overall 10 to 90% risetime can be found by use of Figure 2.46 with $R_2 C_2 / R_1 C_1 = 1.5$, resulting in $t_{10\text{-}90} = 4.25\ R_1 C_1 = 8.5$ nsec. The approximation of eq. (2.82) leads to $t'_{10\text{-}90} = \sqrt{t^2_{10\text{-}90_1} + t^2_{10\text{-}90_2}} = \sqrt{(4.4\ \text{nsec})^2 + (6.6\ \text{nsec})^2} = 7.9$ nsec. Thus, in this example, the square root of the sum of the squares approximation of eq. (2.82) underestimates the 10 to 90% risetime by about 7%. The overall Elmore risetime, from eqs. (2.80), is $t_R = \sqrt{2\pi} \sqrt{(R_1 C_1)^2 + (R_2 C_2)^2} = \sqrt{2\pi} \sqrt{(2\ \text{nsec})^2 + (3\ \text{nsec})^2} = 9.04$ nsec. Thus, in this example, the Elmore risetime overestimates the 10 to 90% risetime by about 6.5%.

Transients of V_R. These are shown in Figure 2.47 for various values of $R_2 C_2 / R_1 C_1$. Note that $R_2 C_2 / R_1 C_1 = \infty$ represents either $R_2 = \infty$ (an open circuit), or $C_2 = \infty$ (a short circuit), or both, and results in a transient of V_R that is identical to V_C of Figure 2.45 with $R_2 C_2 / R_1 C_1 = 0$. Note that when $R_2 C_2 / R_1 C_1$ is larger than about 100—which is usual when V_R is utilized—the rising edge of the transient is not much altered, and for $t \gg R_1 C_1$ voltage V_R can be approximated as $V_R \approx V_0 \exp\left(-t/R_2 C_2\right)$.

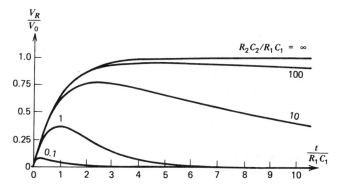

Figure 2.47 Transient response of V_R in the circuit of Figure 2.44 for various values of $R_2 C_2 / R_1 C_1$. The $R_2 C_2 / R_1 C_1 = \infty$ curve represents $R_2 = \infty$ and/or $C_2 = \infty$.

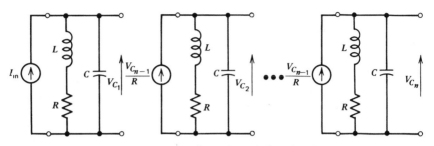

Figure 2.48 Multistage R-L-C circuit consisting of n identical stages.

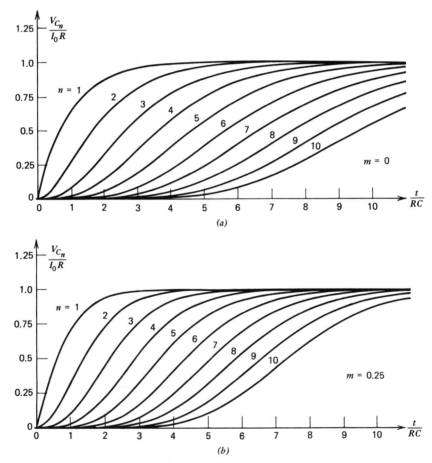

Figure 2.49 Transient response of the circuit of Figure 2.48 for $n = 1$ through 10 stages with various values of $m = L/R^2C$. (a) $m = 0$; (b) $m = 0.25$; (c) $m = 0.41$; (d) $m = 0.5$.

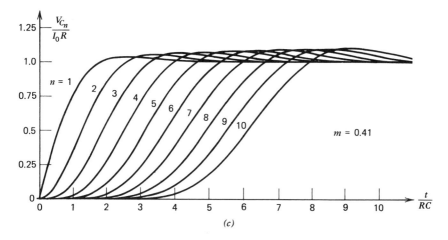

(c)

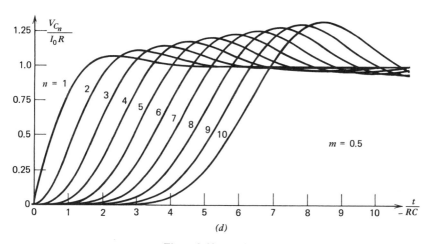

(d)

Figure 2.49 (continued)

2.11.2   Multistage R-L-C Circuits

Figure 2.48 shows a multistage R-L-C circuit consisting of n identical circuits of Figure 2.34. The transient responses for a step current input of

$$I_{in} = I_0 u(t) \tag{2.83}$$

are shown in Figure 2.49 for $n = 1$ through 10 stages with various values of $m = L/R^2 C$. When $m = 0$, each stage simplifies to an R-C circuit; $m = 0.25$

represents the largest value of m that results in monotonic responses; the $m = 0.41$ leads to zero Elmore risetime and to maximally flat frequency response magnitudes, although there is no special effect on the transient response; the $m = 0.5$ was chosen arbitrarily. We can see that when overshoot and ringing are present ($m > 0.25$), these increase rapidly as the number of stages n is increased.

2.11.3 Delay, Risetime, and Overshoot

The overall delay of the 50% point t_{50}, the overall 10 to 90% risetime t_{10-90}, the overall Elmore delay t_D, the overall Elmore risetime t_R, and the overall overshoot ϵ, in the circuit of Figure 2.48 are summarized for $1 \leqslant n \leqslant 10$ in Figure 2.50 through Figure 2.54 with various values of $m = L/R^2 C$ (note that all scales are logarithmic). Also shown are t'_{50} and t'_{10-90} that are defined by

$$t'_{50} = (t_{50})_{n=1}\, n, \tag{2.84}$$

$$t'_{10-90} = (t_{10-90})_{n=1}\, \sqrt{n}. \tag{2.85}$$

Also, the overall Elmore delay and the overall Elmore risetime are given by

$$t_D = (1 - m)\, RCn, \tag{2.86}$$

$$t_R = \sqrt{2\pi}\, \sqrt{1 - m^2 - 2m}\, RC\, \sqrt{n}. \tag{2.87}$$

The $m = 0$ case of Figure 2.50 represents R-C circuits, the $m = 0.25$ case of Figure 2.51 critical damping; neither of these two cases has any overshoot or ringing, thus $\epsilon = 0$. Also for $0 \leqslant m \leqslant 0.25$, reasonable approximations of t_{50} are given by either t_D or by t'_{50}, and reasonable approximations of t_{10-90} are given by either t_R or by t'_{10-90}; further, the approximations by t_D and t_R become more accurate with increasing n.

Figures 2.52 through 2.54 summarize the transients for $m = 0.41, 0.5$, and 1. All these values of m are above 0.25, thus the resulting transients have overshoot and ringing. We can see that reasonable approximations of t_{50} are still given by t'_{50}. The graphs of t_{10-90}, however, do not increase in proportion with $\sqrt{n}$ for large n (an increase in proportion with $\sqrt{n}$ would imply that a graph is parallel with t'_{10-90} which increases as $\sqrt{n}$). Thus, resulting risetimes are *not* equal to the square root of the sum of the squares of the contributing risetimes— as was the case for $m \leqslant 0.25$.

2.11.4 Summary

Transient properties of cascaded circuits have been described in this section. These show that the overall delay of the 50% point t_{50} can be approximated with a maximum error of about 30% by the sum of the contributing delays.

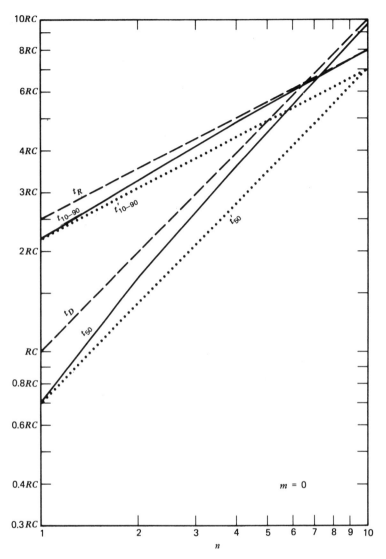

Figure 2.50 Values of t_D, t_R, t_{50}, t_{10-90}, t_{50}', and t_{10-90}' for the transients in the circuit of Figure 2.48 with $m = 0$; $\epsilon = 0$ for this value of m. Solid lines interconnect points for $n = 1$ through 10 and have no significance for noninteger n.

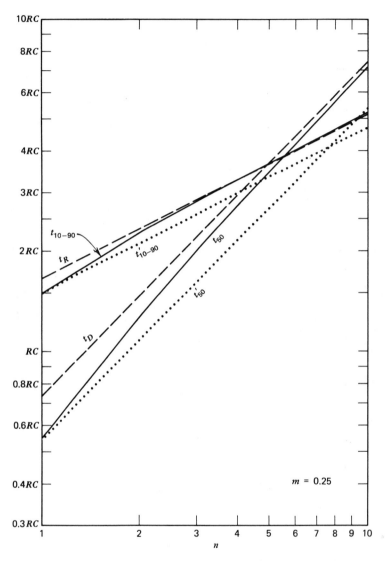

Figure 2.51 Values of t_D, t_R, t_{50}, t_{10-90}, t_{50}', and t_{10-90}' for the transients in the circuit of Figure 2.48 with $m = 0.25$; $\epsilon = 0$ for this value of m. Solid lines interconnect points for $n = 1$ through 10 and have no significance for noninteger n.

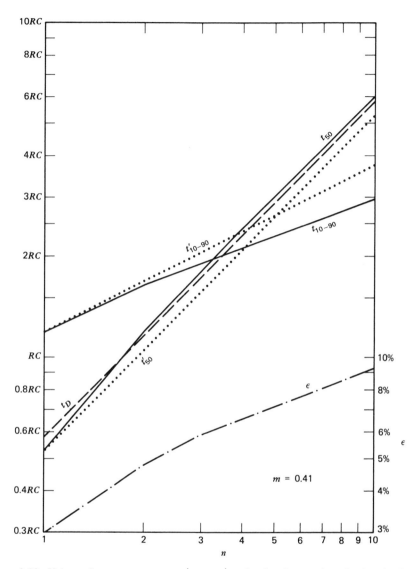

Figure 2.52 Values of t_D, t_{50}, t_{10-90}, t_{50}', t_{10-90}', and ϵ for the transients in the circuit of Figure 2.48 with $m = 0.41$; $t_R = 0$ for this value of m. Solid lines interconnect points for $n = 1$ through 10 and have no significance for noninteger n.

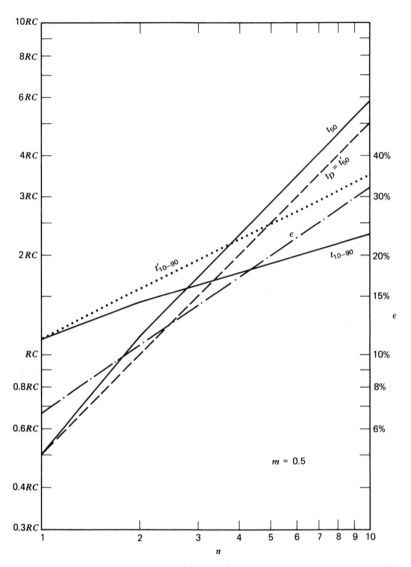

Figure 2.53 Values of t_D, t_{50}, t_{10-90}, t_{50}', t_{10-90}', and ϵ for the transients in the circuit of Figure 2.48 with $m = 0.5$; t_R does not exist for this value of m. Solid lines interconnect points for $n = 1$ through 10 and have no significance for noninteger n.

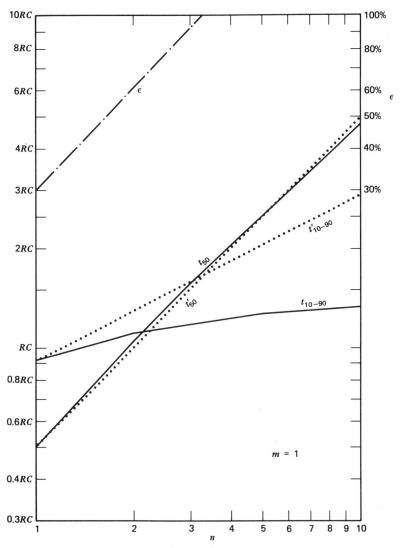

Figure 2.54 Values of t_{50}, t_{10-90}, t_{50}', t_{10-90}', and ϵ for the transients in the circuit of Figure 2.48 with $m = 1$; t_R does not exist, and $t_D = 0$ for this value of m. Solid lines interconnect points for $n = 1$ through 10 and have no significance for noninteger n.

Also, when the transient is monotonic—but not otherwise—the overall 10 to 90% risetime t_{10-90} can be approximated with a maximum error of about 12% by the square root of the sum of the squares of the contributing 10 to 90% risetimes.

When the transient is monotonic, the Elmore delay and the Elmore risetime—which are low frequency characteristics—also provide reasonable approximations of t_{50} and t_{10-90}, respectively, and these approximations improve as the number of stages n is increased.

PROBLEMS

1. Calculate the resistance of 1000 feet of #10 copper wire (diameter $\approx 0.1''$).

2. Verify that the graphs of Figure 2.3 are in agreement with eqs. (2.4) through (2.7).

3. A strobe light used for photography consists of a gas discharge tube that is connected across a capacitor with a capacitance of 500 microfarad = 0.5×10^{-3} farad. The capacitor is initially charged to a voltage of 500 volts, and it is discharged by the tube upon a suitable trigger from the camera. Find the duration of the light flash if, as a crude approximation, the current through the tube is assumed to be a constant 250 amperes until the capacitor is completely discharged.

4. Sketch current $I_L(t)$ in a 1 henry inductor if the voltage in volts across it is given as $V_L(t) = tu(t)$.

5. Demonstrate by an example that the equivalence of Figures 2.8a and 2.9a is not valid when R_G is replaced by a capacitor.

†6. An approximation of the exponential function is obtained from eq. (2.23b) as $e^x \approx (1 + x/2)^2$, another one from eq. (2.25) as $e^x \approx 1 + x + x^2/2$. Sketch these two approximations and compare them with the exact Figure 2.14.

7. Given the rate of change of e^x as $de^x/dx = e^x$, demonstrate by inspection of Figure 2.14 that the rate of change of the function e^{-x} is $de^{-x}/dx = -e^{-x}$. Do not use calculus.

8. The inductance of a strip of conductor with a length l, width w, and thickness t can be approximated as

$$L \approx 0.2 \times 10^{-6} \frac{\text{Henry}}{\text{meter}} l \left(\frac{1}{2} + \ln \frac{2l}{w + t} \right),$$

where l, w, and t are in meters. Find the inductance of a copper strip with a length of 1 foot (≈ 30 cm), a width of 2 inches (≈ 5 cm), and a thickness of 0.03 inch (≈ 0.75 mm).

---

†Optional problems are denoted by †.

9. The current signal shown in the upper graph of Figure 2.3 is used as input current I_{in} in the circuit of Figure 2.16b. Find voltage V_C as function of time if $R = 1$ kΩ, $C = 1$ μF, and if $(V_C)_{t \leqslant 0} = 0$.

†10. The derivation of eqs. (2.33) and (2.34) assumed that $(V_{in})_{t < 0} = 0$. Demonstrate that the two equations are also valid for *any* $(V_{in})_{t < 0}$ as long as $(V_C)_{t=0} = 0$ and $(V_{in})_{t > 0} = V_0$.

11. The pulse V_{in} in Figure 2.20 returns to zero at $t/RC = 8$. Shorten the pulse, without changing its risetime or fall time, such that it returns to zero at $t/RC = 6$. Compute and sketch the resulting V_C.

12. Derive eq. (2.46) by use of eqs. (2.38) and (2.39).

13. Derive eq. (2.50) without using calculus.

14. Find the value of t_r that reduces $V_{L_{max}}$ from 1.9 volt to 1 volt in Example 2.13. What is the resulting $V_{L_{min}}$?

15. Approximate $|V_C|$ of eq. (2.61a) for high frequencies. Specify the range of validity for the approximation in terms of f, f_0, and m. Compare with Figure 2.32.

16. Approximate $|V_C|$ of eq. (2.67) for high frequencies. Specify the range of validity for the approximation in terms of f, f_0, and m. Compare with Figure 2.36.

17. What is the effect of L_p in Example 2.15? Sketch the resulting transient that takes L_p into account. Write an equation for V_{out} as function of time for long times.

†18. Sketch the magnitude and the phase of both V_C and V_R in the circuit of Figure 2.44 with $R_2 C_2/R_1 C_1 = 1$ for a sinewave input of $V_{in} = V_0 \sin 2\pi ft$. Repeat with $R_2 C_2/R_1 C_1 = 10$.

19. Figure 2.55 shows an *operational amplifier* that has an *amplification* $A = V_{out}/(V_p - V_n)$. The frequency response of A is characterized by $A = A_0/$

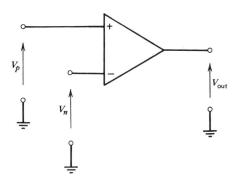

Figure 2.55 An operational amplifier.

$\sqrt{1 + (f/f_0)^2}$ and $\underline{/A}$ = -arctan (f/f_0). Find t_{50}, $t_{10\text{-}90}$, t_D, t_R, and ϵ of output V_{out} in Figure 2.55 as functions of f_0 when $V_n = 0$ and V_p is a step voltage.

†20. Figure 2.56 shows an operational amplifier with *feedback*. It can be shown that when the (-) input of the operational amplifier can be approximated by an open circuit, the resulting *feedback amplification*, $M = V_{\text{out}}/V_{\text{in}}$, can be characterized by $M = M_0/\sqrt{1 + (M_0 f/A_0 f_0)^2}$ and $\underline{/M}$ = -arctan $(M_0 f/A_0 f_0)$ where $M_0 = A_0/[1 + A_0 R_I/(R_I + R_F)]$. Find t_{50}, $t_{10\text{-}90}$, t_D, t_R, and ϵ of output V_{out} in Figure 2.56 as functions of f_0, A_0, and M_0 when V_{in} is a step voltage.

†21. When n stages of the operational amplifier with feedback shown in Figure 2.56 are cascaded, the overall amplification at $f = 0$ is $M_{\text{overall}} = (M_0)^n$ and also $n = (\ln M_{\text{overall}})/(\ln M_0)$, where M_0 is given in the preceding problem.

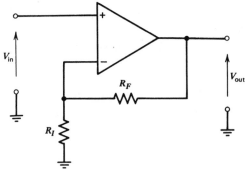

Figure 2.56   An operational amplifier with feedback.

(a) Find the overall Elmore delay $t_{D_{\text{overall}}}$ and the overall Elmore risetime $t_{R_{\text{overall}}}$ for n identical cascaded stages.

(b) Demonstrate that for given f_0 and A_0 and for *a given overall amplification* M_{overall}, $t_{D_{\text{overall}}}$ is minimum when $M_0 = e \approx 2.71$ and $t_{R_{\text{overall}}}$ is minimum when $M_0 = \sqrt{e} \approx 1.65$. *Hint:* Either plot $A_0 f_0 t_{D_{\text{overall}}}$ and $A_0 f_0 t_{R_{\text{overall}}}$ as functions of M_0 and find their minima graphically, or use differential calculus.

REFERENCES

1. A. Barna, *High-Speed Pulse Circuits*, Wiley-Interscience, New York, 1970.

2. A. Barna, *Operational Amplifiers*, Wiley-Interscience, New York, 1971.

3. W. C. Elmore and M. Sands, *Electronics, Experimental Techniques*, McGraw-Hill, New York, 1949.

CHAPTER 3

DIODE CIRCUITS

This chapter describes junction diodes and tunnel diodes, and basic transients in circuits utilizing diodes. It also introduces computer-aided circuit design methods.

3.1 JUNCTION DIODES

Many of today's high speed components are made of *semiconductors* such as germanium, silicon, or gallium-arsenide. To make them suitable for a particular requirement, the properties of these semiconductors are altered by the introduction of *impurities*, which can be *n-type* or *p-type*.* When two adjacent regions of a semiconductor material are dominated, respectively, by n-type and p-type impurities with suitable concentrations, the *junction* between the two regions can be made to exhibit asymmetric electrical characteristics. The resulting device is a *junction diode*.

3.1.1 DC Characteristics

The symbol of a junction diode is shown in Figure 3.1 together with identification of the regions and the voltage and current nomenclature. The current versus voltage characteristics are shown in Figure 3.1*b*. We can see that diode current I_d increases rapidly as diode voltage is increased in the *forward* direction

*Detailed discussions of semiconductor physics and technology can be found in the references at the end of the chapter.

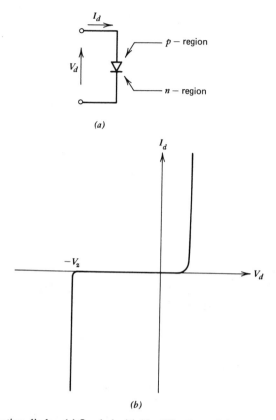

(a)

(b)

Figure 3.1. Junction diode. (a) Symbol with identifications of the n- and p-regions and the voltage and current nomenclature; (b) current versus voltage characteristics.

($V_d > 0$). However, only small currents flow in the *reverse* direction ($V_d < 0$) as long as V_d is above *Zener voltage* $- V_z$. Although diodes used as voltage reference are designed specifically for operation at the Zener voltage (*Zener diodes*), in what follows we assume that V_z is very large and that $-V_z$ is not approached by diode voltage V_d, that is, that $-V_z << -V_d$.

It is evident from Figure 3.1b that I_d is not proportional to V_d, thus Ohm's law of eq. (2.1) is not applicable. The current-voltage relationship has to be defined by a graph as in Figure 3.1b, by an equation, or by a table. The crudest equation that describes the behavior of a junction diode is

$$I_d = I_0 (e^{V_d/V_T} - 1), \tag{3.1a}$$

which can be also written as

$$V_d = V_T \ln \left(1 + \frac{I_d}{I_0}\right). \qquad (3.1b)$$

In eq. (3.1), I_0 is the *saturation current* that is constant for a given diode at a given temperature, and V_T is between 25 mV and 50 mV at room temperature.

Example 3.1 A silicon diode is characterized by $I_0 = 0.01$ nA and $V_T = 25$ mV. If the voltage across the diode is $V_d = 0.5$ V, from eq. (3.1a) the diode current $I_d = 0.01$ nA $(e^{0.5 \text{ V}/0.025 \text{ V}} - 1) = 4.85$ mA. If, however, the current through the diode is $I_d = 50$ mA, from eq. (3.1b) the diode voltage $V_d = 0.025$ V ln $(1 + 50$ mA$/0.01$ nA$) = 0.56$ V. We can also see that I_d reaches $-I_0$ at $V_d = -\infty$ and that I_d can never be more negative than $-I_0$.

Simple as it is, eq. (3.1) provides reasonable approximations for moderate values of I_d, but because of the presence of *series ohmic resistance* it can be grossly inaccurate when I_d is large.

Example 3.2 A silicon diode is characterized by $I_0 = 0.01$ nA and $V_T = 25$ mV. However, the semiconductor material and its connections also have a series ohmic resistance that can be approximated as a constant 5 Ω. If the current through the diode is 50 mA, without the 5 Ω series resistance we would get a voltage drop of $V_d = 0.56$ V (see Example 3.1). By taking the 5 Ω series ohmic resistance into account we get $V_d = 0.56$ V + 50 mA 5 $\Omega = 0.81$ V.

Because of additional effects not included in it, eq. (3.1) is also inaccurate when I_d is very small—which in itself is usually of little consequence in high speed circuits. In germanium and silicon diodes, however, the I_d versus V_d curve is also shifted to the right, thus caution should be exercised when fitting the characteristics of an actual diode by eq. (3.1). In high speed circuits such a fit should be chosen to be reasonably good at the highest current encountered through the diode, even though this will not lead to a good fit for $V_d < 0$. (See Grove in references).

A parameter that is often useful in junction diode circuits is the *incremental resistance* r_i of the diode, which is the slope of the V_d versus I_d curve. Using the property of the exponential function that $d(e^x)/dx = e^x$ we obtain from eq. (3.1) an incremental resistance of

$$r_i = \frac{dV_d}{dI_d} = \frac{1}{(dI_d/dV_d)} = \frac{V_T}{I_0} e^{-V_d/V_T}, \qquad (3.2a)$$

which can be also written as

$$r_i = \frac{V_T}{I_d + I_0}. \qquad (3.2b)$$

Example 3.3 A silicon diode is characterized by I_0 = 0.01 nA and V_T = 25 mV. It is operated at a current of I_d = 5 mA. Thus, according to eq. (3.2b) the incremental resistance r_i = 25 mV/(5 mA + 0.01 nA) = 5 Ω.

Note that r_i is an incremental resistance and *not* a total resistance. Thus an incremental ("small") voltage change dV_d can be found as $dV_d = r_i dI_d$, where dI_d is an incremental ("small") current change; also, dI_d can be found as dV_d/r_i. Such use of r_i is equivalent to replacing the V_d versus I_d curve by a tangent drawn at an operating point (at I_d = 5 mA in Example 3.3 above). We should also note that, in general, $V_d \neq r_i I_d$ and $I_d \neq V_d/r_i$—as can be seen from eqs. (3.1) and (3.2).

3.1.2 Piecewise Linear Approximations

Several piecewise linear approximations of junction diode dc characteristics are shown in Figure 3.2. These approximations are often adequate for crude estimates and we utilize them for such purposes.

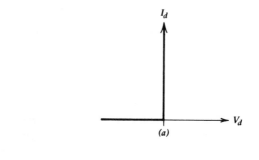

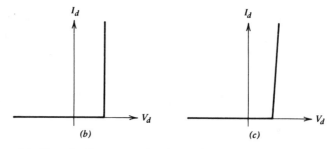

Figure 3.2 Piecewise linear approximations of junction diode dc characteristics.

3.1.3 Stored Charge and Capacitances

The charge stored in a junction diode can be separated into three components, and with each component we can associate a capacitance. In what follows we represent a junction diode by the parallel combination of these three capacitances and the dc I_d versus V_d characteristic.*

One of the three charge components is stored on the *transition capacitance* C_t that is present between the two surfaces of the junction. Another charge component, which is present in the diode as a result of a current flowing through it, is stored on the *diffusion capacitance* C_d. The third charge component is stored on the *stray capacitance* C_s.

Transition Capacitance C_t. This capacitance is a function of the voltage across the diode V_d. While this capacitance and its voltage dependence are significant in many applications, in pulse and digital circuits C_t can be usually lumped into the other two capacitances without incurring drastic errors. For this reason we do not discuss the effects of C_t in detail.

Diffusion Capacitance C_d. The charge stored on this capacitance is present as a result of current I_d flowing through the diode, and we approximate this charge as being proportional to diode current I_d, that is,

$$Q_d = \tau I_d \tag{3.3a}$$

where timeconstant τ is characteristic of the diode. By use of eq. (3.1a), eq. (3.3a) can be also written as

$$Q_d = \tau I_0 (e^{V_d/V_T} - 1). \tag{3.3b}$$

Example 3.4 A silicon diode is characterized by $\tau = 100$ psec, and it is operated at a current of $I_d = 5$ mA. Thus, according to eq. (3.3a) the stored charge in the diode $Q_d = 100$ psec 5 mA = 0.5 pC (picocoulomb).

It is evident from eq. (3.3b) that the stored charge Q_d is not proportional to voltage V_d, hence the linear relationship of eq. (2.4) is not applicable. It is, however, customary to define a *diffusion capacitance* C_d as the slope of the Q_d versus V_d curve. Using the property of the exponential function that $d(e^x)/dx = e^x$ we get

$$C_d = \frac{dQ_d}{dV_d} = \frac{\tau I_0}{V_T} e^{V_d/V_T} \tag{3.4a}$$

*This is a very crude model. For more elaborate treatments see Gibbons, Linvill, and Sze in the references.

which by use of eq. (3.1a) can be also written as

$$C_d = \frac{\tau}{V_T}(I_d + I_0).$$ (3.4b)

Example 3.5 A silicon diode is characterized by $I_0 = 0.01$ nA, $V_T = 25$ mV, and $\tau = 100$ psec, and it is operated at a current of $I_d = 5$ mA. Thus, from eq. (3.4b) the diffusion capacitance $C_d = 100$ psec (5 mA + 0.01 nA)/25 mV = 20 pF.

Note that the combination of eqs. (3.2) and (3.4) leads to

$$r_i C_d = \tau$$ (3.5)

independent of V_d and I_d.

Also note that C_d is an incremental capacitance and *not* a total capacitance. Thus an incremental ("small") change of charge dQ_d can be found as $dQ_d = C_d \, dV_d$, where dV_d is an incremental ("small") voltage change; also dV_d can be found as dQ_d/C_d. Such use of C_d is equivalent to replacing the Q_d versus V_d curve by a tangent drawn at an operating point (at $I_d = 5$ mA in Example 3.5 above). We should also note that, in general, $Q_d \neq C_d V_d$ and $V_d \neq Q_d/C_d$—as can be seen from eqs. (3.3) and (3.4).

Stray Capacitance C_s. This capacitance is approximated as being voltage independent, hence the charge stored on it is simply $C_s V_d$.

3.1.4 Transients in Junction Diode Circuits

The determination of transients in junction diode circuits can become quite difficult even with the use of the simple diode model consisting of a diffusion capacitance, a stray capacitance, and the dc I_d versus V_d characteristic. In general the problem is not analytically tractable, and the use of computer-aided methods are required. Analytical solutions are possible, however, in some simple cases. In what follows here we describe two such cases and return to the general case in the section on computer-aided circuit design.

3.1.4.1 Transients with Diffusion Capacitance

Junction diode transients are described here for the circuit of Figure 3.3a where the only charge storage element is diffusion capacitance C_d.

Turn-on Transient. Consider the circuit of Figure 3.3a with I_g as given in Figure 3.3b. The junction diode is represented by a model consisting of a resistor and

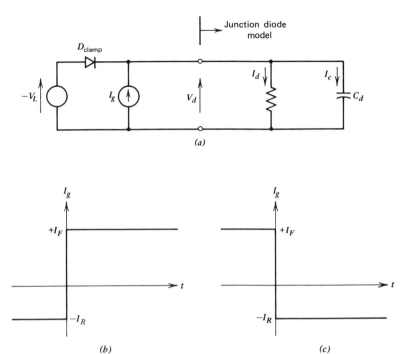

Figure 3.3 Junction diode circuit and input signals. (a) The circuit; (b) input current signal for the turn-on transient; (c) input current signal for the turn-off transient.

a capacitor—both voltage variable. Current I_d is given by eq. (3.1), and the charge stored in C_d is given by eq. (3.3).

Current source I_g can have a value of either $-I_R$ or $+I_F$, where we also assume that $I_F \gg I_0$ and $-I_R \ll -I_0$. The junction diode can not support a current that is more negative than $-I_0$, hence when $I_g = -I_R$ the excess current flows through clamp diode D_{clamp}. For simplicity diode D_{clamp} is characterized by the piecewise linear approximation of Figure 3.2a and by zero stored charge— it turns out that these simplifications are not really necessary, but they make our work easier. We also assume that $-V_L \ll -V_T$. Thus when $I_g = -I_R$ diode voltage $V_d = -V_L$. Subsequent to I_g changing from $-I_R$ to $+I_F$, I_d and V_d start rising, and I_d eventually reaches I_F (at $t = \infty$) and V_d reaches its final value that we designate as forward voltage V_F; by use of eq. (3.1b), $V_F = V_T \ln (1 + I_F/I_0)$.

Next we find the transient of V_d from its initial value $-V_L$ to its final value V_F. As I_g changes from $-I_R$ to $+I_F$ at $t = 0$, V_d starts to rise from $-V_L$ towards $+V_F$. Thus diode D_{clamp} disconnects immediately after $t = 0$, and we are left with a circuit consisting of $I_g = I_F$ and of the junction diode model, that is, for $t > 0$

$$I_F = I_d + I_c. \tag{3.5}$$

From eq. (3.3a) we also have

$$Q_d = \tau I_d. \tag{3.6}$$

Also, since the current is the rate of change of the charge,

$$I_c = \frac{dQ_d}{dt}. \tag{3.7}$$

By use of the fact that τ is constant, the combination of eqs. (3.5), (3.6), and (3.7) leads to

$$I_F = I_d + \frac{dI_d}{d(t/\tau)}. \tag{3.8}$$

We seek solution of eq. (3.8) in the form

$$I_d = Ae^{-t/\tau} + B \tag{3.9}$$

where A and B remain to be found. By using the property of the exponential function that $d(e^{-x})/dx = -e^{-x}$, we get from eq. (3.9):

$$\frac{dI_d}{d(t/\tau)} = -Ae^{-t/\tau}. \tag{3.10}$$

Substitution of eqs. (3.9) and (3.10) into eq. (3.8) results in $B = I_F$, hence we can rewrite eq. (3.9) as

$$I_d = Ae^{-t/\tau} + I_F, \tag{3.11}$$

where A still remains to be found. Now we use the fact that $V_d = -V_L$ at $t = 0$, and by use of eq. (3.1) we write

$$(I_d)_{t=0} = I_0(e^{-V_L/V_T} - 1). \tag{3.12}$$

However, by applying eq. (3.11) to $t = 0$

$$(I_d)_{t=0} = A + I_F. \tag{3.13}$$

The combination of eqs. (3.12) and (3.13) leads to $A = I_0 [\exp(-V_L/V_T) - 1] - I_F$, whence eq. (3.11) becomes

$$I_d = I_F \left\{ 1 - \left[1 - \frac{I_0}{I_F} (e^{-V_L/V_T} - 1) \right] e^{-t/\tau} \right\}. \tag{3.14}$$

Also, by the use of eq. (3.1b) and $V_F = V_T \ln(1 + I_F/I_0)$ we get from eq. (3.14):

$$V_d = V_F + V_T \ln \left(1 - e^{-t/\tau} + \frac{e^{-V_L/V_T}}{1 + (I_F/I_0)} e^{-t/\tau} \right). \tag{3.15}$$

Equations (3.14) and (3.15) describe the turn-on transient of the junction diode. A check shows that at $t = 0$ eq. (3.14) indeed reverts to eq. (3.12), and at $t = \infty$ it leads to $I_d = I_F$. Also at $t = 0$ eq. (3.15) reduces to $(V_d)_{t=0} = -V_L$ and at $t = \infty$ to $(V_d)_{t=\infty} = V_F$.

Note that since $V_L \gg V_T$ and $I_F \gg I_0$, the last term in eq. (3.15) is small and can be neglected for all times except *at* $t = 0$. Thus we can write

$$V_d \approx V_F + V_T \ln(1 - e^{-t/\tau}), \text{ for } t > 0. \tag{3.16}$$

At $t = 0$ eq. (3.16) would lead to $(V_d)_{t=0} = -\infty$, thus in effect eq. (3.16) approximates $-V_L$ by $-\infty$. Also, as expected, at $t = \infty$ eq. (3.16) leads to $(V_d)_{t=\infty} = V_F$. For times in between, V_d is a function of V_F, V_T, τ, and of time t. The shape of V_d as function of time, however, is independent of V_F, V_d, and τ, thus we can characterize V_d by rewriting eq. (3.16) as

$$\frac{V_d - V_F}{V_T} = \ln(1 - e^{-t/\tau}), \tag{3.17}$$

and plotting it as shown in Figure 3.4. Thus when V_F, V_T, and τ are given, the transient can be obtained by appropriate relabeling of the axes in Figure 3.4.

Example 3.6 In Figure 3.3 the junction diode is characterized by $I_0 = 0.01$ nA, $V_T = 25$ mV, and $\tau = 100$ psec. Current source I_g is characterized by $I_R = 10$ mA and $I_F = 35$ mA; $-V_L = -5$ V. By use of eq. (3.1b), $V_F =$

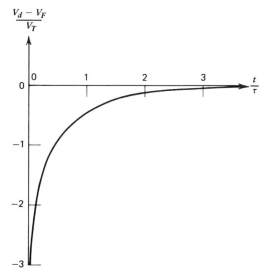

Figure 3.4 Turn-on transient in the circuit of Figure 3.3a with the input current of Figure 3.3b.

25 mV ln $(1 + 35$ mA/0.01 nA$) = 25$ mV $\times$ 22 = 0.55 V. Hence on the vertical axis $(V_d - V_F)/V_T = 0$ corresponds to and should be relabeled as $V_d = 0.550$ V; $(V_d - V_F)/V_T = -1$ as $V_d = 0.525$ V; and so on. On the horizontal axis $t/\tau = 0$ corresponds to and should be relabeled as $t = 0$; $t/\tau = 1$ as $t = 100$ psec; and so on. Thus, for example, at $t = 80$ psec we get $(V_d)_{t=80 \text{ psec}} = 0.535$ V.

We can see that the turn-on transient of the junction diode is quite fast: at $t = 2\tau$ voltage V_d is already within 0.15 V_T of its final value V_F. The reason for this speed is that we neglected stray capacitance C_S as well as the transition capacitance. These often dominate the turn-on transient when τ is small.

Example 3.7 We modify Example 3.6 by including in parallel with the junction diode a stray capacitance of $C_S = 1$ pF. If only C_S were present then the full current I_F would be available to charge C_S, and $V_d = 0.535$ V would be reached at a time $t = C_S(V_d + V_L)/I_F = 1$ pF $(0.535$ V $+ 5$ V$)/35$ mA $= 160$ psec. When the junction diode is present in addition to C_S, only part of I_F is available to charge C_S, and $V_d = 0.535$ V will be reached at some time $t > 160$ psec—which is not anywhere near the 80 psec that was the case when C_S was neglected.

Turn-off Transient. Here we use the circuit of Figure 3.3a with I_g as given in Figure 3.3c. Again we assume $I_F \gg I_0$, $-I_R \ll -I_0$, and $-V_L \ll -V_T$, as well as a clamp diode D_{clamp} with zero stored charge. Now the transient of V_d starts at forward voltage V_F, and at $t = 0$ it starts dropping towards $V_d = -\infty$. However, for $t > 0$ we now have

$$-I_R = I_d + I_c \qquad (3.18)$$

as well as eqs. (3.6) and (3.7), whence we get

$$-I_R = I_d + \frac{dI_d}{d(t/\tau)} . \qquad (3.19)$$

As before, we seek solution of eq. (3.19) in the form

$$I_d = Ae^{-t/\tau} + B \qquad (3.20)$$

where A and B remain to be found. By using the property of the exponential function that $d(e^{-x})/dx = -e^{-x}$, we get from eq. (3.20)

$$\frac{dI_d}{d(t/\tau)} = -Ae^{-t/\tau}. \qquad (3.21)$$

Substitution of eqs. (3.20) and (3.21) into eq. (3.19) results in $B = -I_R$, hence we can rewrite eq. (3.20) as

$$I_d = Ae^{-t/\tau} - I_R \qquad (3.22)$$

where A still remains to be found. Now we use the fact that at $t = 0$, $I_d = I_F$ and that, from eq. (3.22),

$$(I_d)_{t=0} = A - I_R; \qquad (3.23)$$

whence $A = (I_d)_{t=0} + I_R = I_F + I_R$ and thus

$$I_d = (I_F + I_R) e^{-t/\tau} - I_R. \qquad (3.24)$$

Also, by the use of eq. (3.1a) and $V_F = V_T \ln (1 + I_F/I_0)$,

$$V_d = V_F + V_T \ln \left[1 - \frac{I_F + I_R}{I_F + I_0} (1 - e^{-t/\tau}) \right]. \qquad (3.25)$$

It is customary to define a *storage time* t_s as the time when V_d of eq. (3.25) reaches zero. This can be obtained from eq. (3.25) as

$$t_s = \tau \ln \left(1 + \frac{I_F}{I_R} \right). \qquad (3.26)$$

Equation (3.26) is illustrated in Figure 3.5.

Example 3.8 In Figure 3.3, $I_R = 5$ mA, $I_F = 50$ mA, and the junction diode is characterized by $\tau = 100$ psec. Thus, from eq. (3.26) the storage time is $t_s = 100$ psec ln (1 + 50 mA/5 mA) = 240 psec.

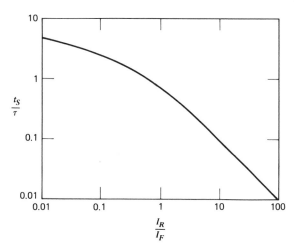

Figure 3.5 Storage time t_s for the turn-off transient in the circuit of Figure 3.3a with the input current of Figure 3.3c.

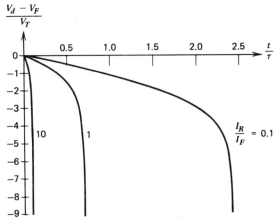

Figure 3.6 Turn-off transients in the circuit of Figure 3.3a with the input current of Figure 3.3c and with I_R/I_F = 0.1, 1, and 10.

Now we take into account that $I_F \gg I_0$ and $I_R \gg I_0$, whence we get from eq. (3.25) the approximation

$$V_d \approx V_F + V_T \ln\left[-\frac{I_R}{I_F} + \left(1 + \frac{I_R}{I_F}\right)e^{-t/\tau}\right]. \tag{3.27}$$

The $(V_d - V_F)/V_T$ resulting from eq. (3.27) is illustrated in Figure 3.6 for various values of I_R/I_F. The initial slow decrease of V_d is followed by a fast drop, but—as with the turn-on transient—in reality the steep slopes are moderated by the presence of stray capacitance. Also, as expected, the transients become faster as I_R/I_F is increased.

Note that eqs. (3.25) and (3.27) are valid only as long as $V_d > -V_L$, since at $V_d = -V_L$ clamp diode D_{clamp} turns on, and eqs. (3.24) and (3.25) cease to be valid. When $-V_L \ll -V_T$, this time can be obtained from eq. (3.25) as

$$t_{-\infty} = \tau \ln\left(1 + \frac{I_F + I_0}{I_R - I_0}\right). \tag{3.28}$$

We can see that this is quite close to t_s, and it becomes equal to it when $I_F \gg I_0$ and $I_R \gg I_0$.

3.1.4.2 Transients with Stray Capacitance

Junction diode transients are given here for the circuit of Figure 3.7 where the only charge storage element is stray capacitance C_S. As before, we assume that clamp diode D_{clamp} is characterized by the piecewise linear approximation of

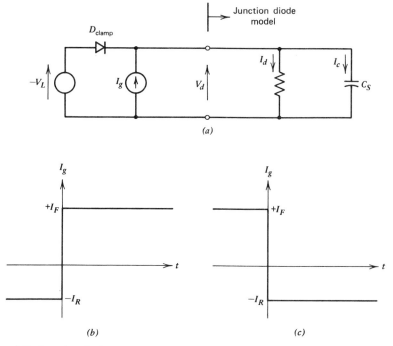

Figure 3.7 Junction diode circuit and input signals. (a) The circuit; (b) input current signal for the turn-on transient; (c) input current signal for the turn-off transient.

Figure 3.2a and by zero stored charge. We also assume that the dc characteristics of the diode are described by eq. (3.1) and that $I_F \gg I_0$ and $I_R \gg I_0$.

Turn-on Transient. It can be shown that with the above assumptions the transient in the circuit of Figure 3.7a with the I_g of Figure 3.7b can be written as

$$V_d = V_F - V_T \ln \left\{ 1 + \exp \left[- \frac{I_F}{C_S V_T} (t - t_1) \right] \right\} \qquad (3.29a)$$

where V_F is the forward voltage

$$V_F = V_T \ln \left(1 + \frac{I_F}{I_0} \right) \qquad (3.29b)$$

and

$$t_1 = (V_L + V_F) \frac{C_S}{I_F} \qquad (3.29c)$$

is the time when voltage V_F would be reached if I_d in Figure 3.7a were zero,

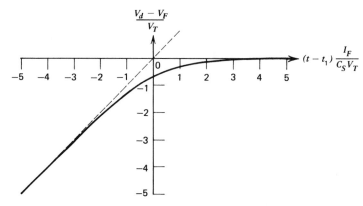

Figure 3.8 Turn-on transient in the circuit of Figure 3.7a with the input current of Figure 3.7b.

that is, if we had only C_S to charge. The $(V_d - V_F)/V_T$ resulting from eq. (3.29) is plotted in Figure 3.8 as a function of $(t - t_1) I_F/C_S V_T$. Figure 3.8 also shows by broken line the transient that would result if I_d in Figure 3.7a were zero.

Example 3.9 In the circuit of Figure 3.7 the junction diode is characterized by $I_0 = 0.01$ nA and $V_T = 25$ mV. Current source I_g is characterized by $I_R = 5$ mA and $I_F = 5$ mA; also $V_L = 5$ V and $C_S = 1$ pF. By use of eq. (3.1b) we get $V_F = 25$ mV ln $(1 + 5$ mA$/0.01$ nA$) = 0.5$ V. Thus, from eq. (3.29c), $t_1 = (V_L + V_F)C_S/I_F = (5$ V $+ 0.5$ V$)$ 1 pF$/5$ mA $= 1.1$ nsec. Hence, on the vertical axis of Figure 3.8, $(V_d - V_F)/V_T = 0$ corresponds to $V_d = 0.5$ V, $(V_d - V_F)/V_T = -1$ to $V_d = 0.475$ V, and so on. On the horizontal axis $(t - t_1)I_F/C_S V_T = 0$ corresponds to $t = t_1 = 1.1$ nsec. Also, since $C_S V_T/I_F = 1$ pF 25 mV$/5$ mA $= 5$ psec, $(t - t_1)I_F/C_S V_T = 1$ corresponds to $t = 1.105$ nsec, $(t - t_1)I_F/C_S V_T = 2$ to $t = 1.110$ nsec, and so on. Thus, for example, at $t = 1.1$ nsec we get a $V_d = 0.4825$ V and at $t = 1.105$ nsec a $V_d = 0.4922$ V.

Turn-off Transient. It can be shown that with the same assumptions as above, the transient in the circuit of Figure 3.7a with the I_g of Figure 3.7c can be written as

$$V_d = V_F - V_T \ln\left[-\frac{I_F}{I_R} + \left(1 + \frac{I_F}{I_R}\right)\exp\left(\frac{I_R}{C_S V_T}\, t\right)\right]. \qquad (3.30)$$

The $(V_d - V_F)/V_T$ resulting from eq. (3.30) is illustrated in Figure 3.9 for various values of I_F/I_R. Also, for times $t \gg C_S V_T/I_R$, V_d approaches

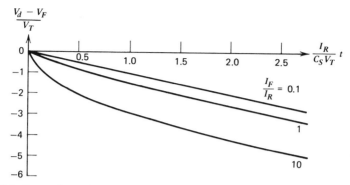

Figure 3.9 Turn-off transients in the circuit of Figure 3.7a with the input current of Figure 3.7c.

$$V_d = V_F - \frac{I_R t}{C_S} - V_T \ln\left(1 + \frac{I_F}{I_R}\right); \quad t \gg \frac{C_S V_T}{I_R}. \tag{3.31}$$

It can be also shown that the $V_d = 0$ points on the transients of Figure 3.9 are reached at times t_s given by

$$t_s = \left[\frac{V_F}{V_T} - \ln\left(1 + \frac{I_F}{I_R}\right)\right]\frac{C_S V_T}{I_R}. \tag{3.32}$$

When I_F is not much larger than I_R, eq. (3.32) can be approximated as

$$t_s \approx \frac{C_S V_F}{I_R}, \tag{3.33}$$

that is, this approximation assumes that the full current I_R is available for discharging C_S and that the current flowing as I_d is negligible during the transient.

Example 3.10 In the circuit of Figure 3.7 the junction diode is characterized by $I_0 = 0.01$ nA and $V_T = 25$ mV. Current source I_g is characterized by $I_R = 1$ mA and $I_F = 5$ mA; $C_S = 1$ pF. Thus $V_F = 0.5$ V and from eq. (3.32) $t_s = 96.6$ psec. The less accurate eq. (3.33) results in $t_s \approx 100$ psec.

3.2 COMPUTER-AIDED CIRCUIT DESIGN

In this section we describe a few simple graphical and numerical methods that are applicable even when solution by analytical means is not feasible. First we look at methods applicable to dc problems and then at one applicable to transients.

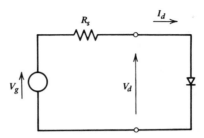

Figure 3.10   Junction diode circuit with series resistance.

3.2.1 DC Methods

Consider the circuit of Figure 3.10. Voltage V_g has a constant positive value, R_s is fixed, and the junction diode is described by

$$I_d = I_0 \, (e^{V_d/V_T} - 1) \tag{3.34a}$$

or by

$$V_d = V_T \ln \left(1 + \frac{I_d}{I_0}\right). \tag{3.34b}$$

The values of V_d and I_d are also connected by way of R_s and V_g as

$$I_d = \frac{V_g - V_d}{R_s} \tag{3.35a}$$

or as

$$V_d = V_g - R_s I_d. \tag{3.35b}$$

The simultaneous solution of eqs. (3.34) and (3.35) would yield V_d and I_d. Unfortunately, neither V_d nor I_d can be explicitly obtained from eqs. (3.34) and (3.35), and they have to be found by graphical or numerical means.

3.2.1.1 Graphical Methods

Here we rephrase the problem as follows: The junction diode on the right side of the circuit imposes the condition on I_d given by eq. (3.34a); this current is imposed by the diode, and hence we denote it by I_{d_D}:

$$I_{d_D} = I_0 \, (e^{V_d/V_T} - 1). \tag{3.36}$$

Similarly, V_g and R_s on the left side of the circuit impose the condition on I_d given by eq. (3.35a); this current is imposed by way of R_s, and hence we denote it by I_{d_R}:

$$I_{d_R} = \frac{V_g - V_d}{R_s}. \tag{3.37}$$

The solution is obtained by plotting eqs. (3.36) and (3.37) and finding the point where

$$I_{d_D} = I_{d_R}. \tag{3.38}$$

Example 3.11 Equations (3.36) and (3.37) with $V_g = 1$ V, $R_s = 20$ Ω, $I_0 = 0.01$ nA, and $V_T = 25$ mV are shown in Figure 3.11a and with expanded scales in Figure 3.11b. We can see that I_{d_D} and I_{d_R} intersect at $V_d \approx 0.539$ V, and at that point the current is $I_d \approx 23$ mA.

Such graphical solutions require the presence of the human element for finding the intersection of the two curves—even if these are displayed on a

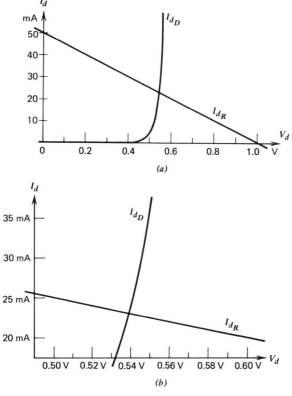

(a)

(b)

Figure 3.11 Graphical solution of V_d and I_d in the circuit of Figure 3.10 with $V_g = 1$ V, $R_g = 20$ Ω, $I_0 = 0.01$ nA, and $V_T = 25$ mV. (a) Overall conditions. (b) Enlarged detail in the vicinity of the intersection.

computer terminal. The numerical methods that follow require only a correct initial problem definition whence a digital computer can proceed to find the solution.

3.2.1.2 Halving the Difference

This numerical method can find the intersection of two curves with a prescribed accuracy, provided that two criteria are met. According to the first criterion, the method is applicable only within a region where both curves are continuous, single-valued, and monotonic, and they intersect at one point and only one point.

Example 3.12 In Figure 3.11b the first criterion is satisfied in the region given by $0.5 \text{ V} \leqslant V_d \leqslant 0.6 \text{ V}$ or by $20 \text{ mA} \leqslant I_d \leqslant 25 \text{ mA}$. Regions that do not satisfy the first criterion will be seen in Section 3.3 on tunnel diodes.

The second criterion is that we must be able to explicitly express both curves as functions of the same variable.

Example 3.13 Both curves shown in Figure 3.11b can be explicitly expressed as functions of V_d by use of eqs. (3.34a) and (3.35a), thus they satisfy the second criterion. Actually, they can be explicitly expressed also as functions of I_d by use of eqs. (3.34b) and (3.35b)–which leads to some flexibility.

Figure 3.12 shows a *flowchart* of a computer program that can find the intersection of the two curves given by eqs. (3.34b) and (3.35b); however, these equations have been rewritten as

$$V_{d_D} = V_T \ln \left(1 + \frac{I_d}{I_0} \right) \tag{3.39}$$

and

$$V_{d_R} = V_g - R_s I_d. \tag{3.40}$$

The use of the flowchart of Figure 3.12 is illustrated by the example that follows.

Example 3.14 This example uses the flowchart of Figure 3.12 for finding the intersection of the two curves in Figure 3.11b. We use eqs. (3.39) and (3.40) with values of $V_g = 1 \text{ V}, R_s = 20 \text{ }\Omega, I_0 = 0.01 \text{ nA}$, and $V_T = 25 \text{ mV}$. The criteria for the use of the halving-the-difference method are satisfied in the region defined in Figure 3.11b by a minimum I_d of $I_{d_{\min}} = 20 \text{ mA}$

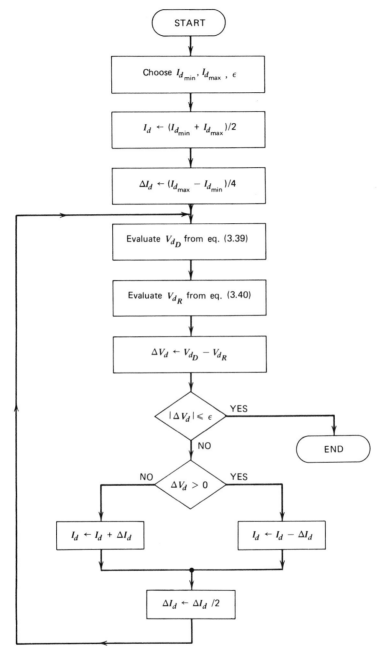

Figure 3.12 Flowchart of the halving-the-difference computer program for finding V_d and I_d in the circuit of Figure 3.10.

and a maximum I_d of $I_{d_{max}}$ = 30 mA. We also set an arbitrary accuracy of ϵ = 0.2 mV to be attained by V_d.

The first block after START in Figure 3.12 is already completed. The second block *assigns* by the horizontal arrow to I_d a value that is halfway between $I_{d_{min}}$ and $I_{d_{max}}$, that is, a value of $(I_{d_{min}} + I_{d_{max}})/2 = (20$ mA + 30 mA$)/2$ = 25 mA–this will be our first guess for the correct value of I_d. The third block assigns to *current difference* ΔI_d an initial value that is one-fourth the size of our region in the direction of I_d, that is, ΔI_d becomes $(I_{d_{max}} - I_{d_{min}})/4 = (30$ mA - 20 mA$)/4$ = 2.5 mA–we later change I_d by this amount to obtain our second guess.

Proceeding down the flowchart, the next three steps evaluate V_{d_D}, V_{d_R}, and their difference ΔV_d, resulting in V_{d_D} = 0.541 V, V_{d_R} = 0.5 V, and ΔV_d = 0.041 V.

From the second criterion we know that the two curves intersect somewhere between $I_{d_{min}}$ = 20 mA and $I_{d_{max}}$ = 30 mA. Our first guess resulted in a voltage difference of ΔV_d = 0.041 V and hence also in $|\Delta V_d|$ = 0.041 V, which is larger than the maximum permissible ϵ = 0.2 mV. Thus we leave the diamond-shaped block containing $|\Delta V_d| \leq \epsilon$ by way of its NO exit; this block will later signal when the desired accuracy is reached by a routing by way of its YES exit to END. Our initial guess of I_d = 25 mA, however, led to an unacceptable error.

We track our progress in Figure 3.13, where we can see our first guess

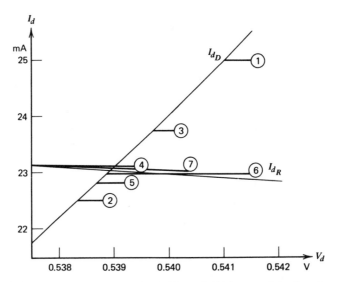

Figure 3.13 Solution for the intersection in Figure 3.11b by use of the flowchart shown in Figure 3.12. Circled numbers refer to line numbers in Table 3.1.

Table 3.1

Line Number	I_d (mA)	ΔI_d (mA)	V_{d_D} (V)	V_{d_R} (V)	ΔV_d (V)	$|\Delta V_d| \leqslant \epsilon$	$\Delta V_d > 0$
①	25.0	2.5	0.541	0.50	0.041	NO	YES
②	22.5	1.25	0.538	0.55	-0.012	NO	NO
③	23.75	0.625	0.540	0.525	0.015	NO	YES
④	23.125	0.3125	0.5390	0.5375	0.0015	NO	YES
⑤	22.8125	0.15625	0.5387	0.5438	-0.005	NO	NO
⑥	22.9688	0.07813	0.5389	0.5406	-0.0017	NO	NO
⑦	23.0469	0.03906	0.53896	0.53906	-0.0001	YES	

marked by ①. We also tabulate our progress in Table 3.1: thus far we just passed the $|\Delta V_d| \leqslant \epsilon$ column with a NO entry in line ①.

Now we make a second guess for I_d. The first guess of $I_d = 25$ mA resulted in $V_{d_D} > V_{d_R}$, that is, in $\Delta V_d > 0$, indicating that the value of I_d is too high. Thus our second guess should be lower than the first one: somewhere between 20 mA and 25 mA—we take 25 mA - ΔI_d = 25 mA - 2.5 mA = 22.5 mA. This process is outlined by the diamond shaped block of $\Delta V_d > 0$ and by its YES exit leading to the new value of 22.5 mA for I_d. Next we halve ΔI_d for later use resulting in $\Delta I_d = 1.25$ mA, and proceed up and back into the flowchart evaluating V_{d_D}, V_{d_R}, and ΔV_d as shown in line ② of Table 3.1. Since we still have $|\Delta V_d| \not\leqslant \epsilon$, we get to the $\Delta V_d > 0$ block which we now exit on its NO exit, add 1.25 mA to I_d, change ΔI_d to 0.625 mA, and keep going as shown in Figure 3.13 and Table 3.1.

We can see that once the flowchart of Figure 3.12 is set up—and this involves an evaluation of the problem by defining a suitable region—the solution proceeds routinely until the desired accuracy is attained.

3.2.1.3 Spiraling in

As with the preceding method, the spiraling-in method requires that a region be found where both functions are continuous, single-valued, and monotonic, and they intersect at one point and only at one point. Here, however, we express each curve as a function of a different variable: for the circuit of Figure 3.10 we use eqs. (3.34b) and (3.35a). The flowchart for this method is shown in Figure 3.14, and its use is illustrated by the example that follows.

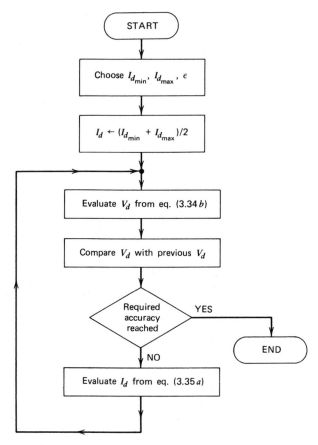

Figure 3.14 Flowchart of the spiraling-in computer program for finding V_d and I_d in the circuit of Figure 3.10.

Example 3.15 This example uses the flowchart of Figure 3.14 for finding the intersection of the two curves in Figure 3.11b. We use eqs. (3.34b) and (3.35a) with values of $V_g = 1$ V, $R_s = 20$ Ω, $I_0 = 0.01$ nA, and $V_T = 25$ mV. We also set an accuracy of $\epsilon = 0.2$ mV to be attained by V_d.

In the first block after START in Figure 3.14 we choose $I_{d_{\min}} = 20$ mA and $I_{d_{\max}} = 30$ mA; we already have $\epsilon = 0.2$ mV. The second block provides the first guess for I_d as $(I_{d_{\min}} + I_{d_{\max}})/2 = (20 \text{ mA} + 30 \text{ mA})/2 = 25$ mA. The next block computes the value of V_d from eq. (3.34b) resulting in 0.541 V. The diamond-shaped block is now reached: it is unlikely that our first guess satisfies the accuracy requirement thus we proceed through the NO exit to the last block.

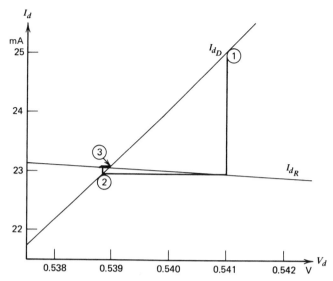

Figure 3.15 Solution for the intersection of Figure 3.11*b* by use of the flowchart shown in Figure 3.14. Circled numbers refer to line numbers in Table 3.2.

We track our progress in Figure 3.15 where we can see our first guess marked by ①. We also tabulate our progress in Table 3.2: thus far we completed line ①.

The last block in Figure 3.14 provides the second guess for I_d and we proceed up and back into the flowchart. We can see in Table 3.2 that this method attains the required accuracy of $\epsilon = 0.2$ mV in V_d in three lines—as compared to the seven lines in Table 3.1. Figure 3.15 shows that the intersection of the two curves is approached along a spiral.

Often, as in Example 3.15, convergence of the spiraling-in method is very fast. In other cases, however, the method diverges and can not be used, as in Problem 3.12 at the end of the chapter.

Table 3.2

Line Number	I_d (mA)	V_d (V)	Accuracy Attained
①	25.0	0.541	NO
②	22.9506	0.53885	NO
③	23.0575	0.53897	YES

3.2.2 Transient Methods

Consider the junction diode circuit of Figure 3.16a. It is similar to the circuit of Figure 3.10 except that a constant capacitance C has been added in parallel to the diode. We are interested in the turn-off transient, that is, we want to determine V_d as function of time for input voltage V_g as shown in Figure 3.16b. However, unlike in the circuit of Figure 3.7, the transients in the circuit of Figure 3.16a are not analytically tractable, and a computer-aided method will be used.

Several computer-aided methods have been developed for the analysis of transients in nonlinear circuits. In what follows here, we illustrate the use of one of the simplest methods, the *forward-Euler method*. The same method is also used later for finding transients in tunnel-diode and transistor circuits.

First we have to find the initial value of V_d in Figure 3.16a, which value we denote by V_{d0}. This can be handled by one of the dc methods described above. Thus, for example, when $V_F = 1$ V, $R = 20$ Ω, and the diode is characterized by eq. (3.34a) with $I_0 = 0.01$ nA and $V_T = 25$ mV, we get from Table 3.2 a $V_{d0} \approx 0.5390$ V.

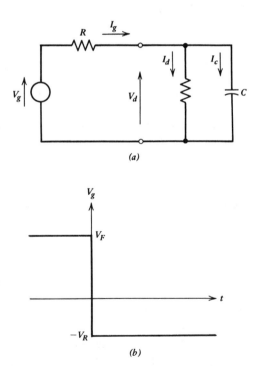

(a)

(b)

Figure 3.16 Junction diode circuit with series resistance and stray capacitance. (a) The circuit; (b) input voltage signal.

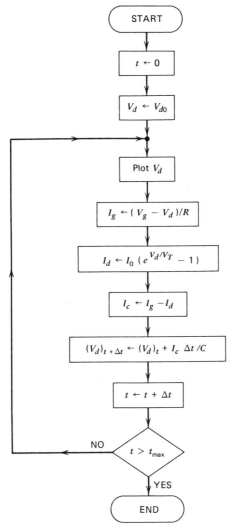

Figure 3.17 Flowchart of the forward-Euler computer program for the turn-off transient in the circuit of Figure 3.16.

In order to determine the transient, we write in the circuit of Figure 3.16 for $t > 0$:

$$I_g = \frac{V_g - V_d}{R}, \tag{3.41}$$

$$I_d = I_0 (e^{V_d/V_T} - 1), \tag{3.42}$$

$$I_c = I_g - I_d. \tag{3.43}$$

Further, based on Section 2.2 we can write

$$(V_d)_{t+\Delta t} \approx (V_d)_t + \frac{I_c \Delta t}{C}, \tag{3.44}$$

where we choose a time interval Δt that is short to permit the approximation of I_c by its value at time t throughout the time interval lasting from time t to time $t + \Delta t$.

Next we would like to solve eqs. (3.41) through (3.44) for V_d as function of time. However, this is not possible analytically, and we follow an iterative numerical procedure whereby we repeatedly evaluate eqs. (3.41) through (3.44) at times 0, Δt, 2 Δt, 3 Δt, and so on, up to a time t_{max}. The procedure is outlined in the flowchart of Figure 3.17. Note the meaning of the ← symbol: "$x \leftarrow y$" stands for "the value of y is assigned to x." The use of the flowchart of Figure 3.17 is illustrated in the example that follows.

Example 3.16 In the circuit of Figure 3.16, $V_F = 1$ V, $-V_R = -0.1$ V, $R = 20\ \Omega$, $C = 50$ pF, and in eq. (3.42) $I_0 = 0.01$ nA and $V_T = 25$ mV. As a crude first choice we use $\Delta t = 0.25\ RC = 0.25$ nsec. We track our progress in Table 3.3; the resulting V_d is also plotted as the lower graph in Figure 3.18. We stop at 1 nsec in Table 3.3, but in Figure 3.18 V_d is plotted up to a time of $t_{max} = 5.5$ nsec. Figure 3.18 also shows the turn-off transient computed with a much smaller Δt, namely with $\Delta t = 0.01\ RC = 0.01$ nsec; it can be also shown that further reduction of Δt changes V_d by less than the line width in Figure 3.18. Thus, if we accept the line width as a criterion of accuracy, the $\Delta t/RC = 0.01$ curve in Figure 3.18 provides the turn-off transient of V_d.

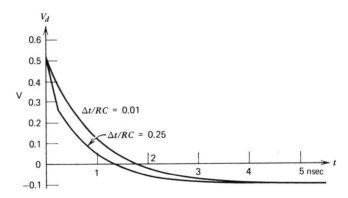

Figure 3.18 Turn-off transients of V_d in Example 3.16.

Table 3.3

t (nsec)	$(V_d)_t$ (V)	I_g (mA)	I_d (mA)	I_c (mA)	$\dfrac{I_c \Delta t}{C}$ (V)	$(V_d)_{t+\Delta t}$ (V)
0.0	0.5390	−31.95	23.05	−55.00	−0.2750	0.2640
0.25	0.2640	−18.20	0.0003	−18.20	−0.0910	0.1730
0.50	0.1730	−13.65	≈0	−13.65	−0.0683	0.1047
0.75	0.1047	−10.24	≈0	−10.24	−0.0512	0.0535
1.00	0.0535					

†3.3 TUNNEL DIODES

The tunnel diode is a junction diode with impurity concentrations chosen such that it exhibits an S-shaped current versus voltage characteristic. This section describes dc and transient characteristics of tunnel diodes and of simple tunnel-diode circuits.

3.3.1 DC Characteristics

The dc current versus voltage characteristic of a tunnel diode can be approximated as

$$I_{td} = I_d + I_t \tag{3.45a}$$

where

$$I_d = I_0 (e^{V_d/V_T} - 1) \tag{3.45b}$$

and I_t is the *tunneling current*

$$I_t = I_p \frac{V_d}{V_p} e^{(1 - V_d/V_p)}. \tag{3.45c}$$

The characteristics described by eqs. (3.45) are illustrated in Figure 3.19 for a germanium diode with $V_p = 3 V_T$ and $I_0 = 5 \times 10^{-10} I_p$. Voltage V_p is the *peak voltage*, I_p the *peak current*, V_v the *valley voltage*, I_v the *valley current*, and V_f the *forward voltage*.

We can see that between voltages of V_p and V_v the slope of the current versus voltage characteristic is negative and the tunnel diode exhibits a *negative resistance*. It can be also shown that in this region the curve is steepest and the magnitude of the negative resistance is largest at approximately $V_d = 2 V_p$, where it has a value of

$$r_i = -e V_p/I_p = -2.71 V_p/I_p. \tag{3.46}$$

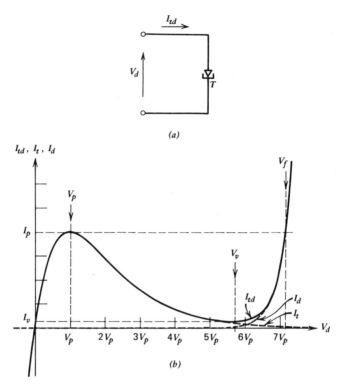

Figure 3.19   Tunnel diode. (a) Symbol with voltage and current nomenclature; (b) current versus voltage characteristics (solid line) and current components (broken lines).

Example 3.17   A germanium tunnel diode is characterized by $V_T = 25$ mV, $V_p = 3\ V_T = 75$ mV, $I_p = 10$ mA, and $I_0 = 5 \times 10^{-10}\ I_p = 5$ pA. Thus Figure 3.19b is applicable with $I_p = 10$ mA and $V_p = 75$ mV. The valley voltage is $V_v = 5.7\ V_p = 427$ mV, the valley current is $I_v = 0.065\ I_p = 0.65$ mA, and the negative resistance has the largest magnitude at $V_d = 2\ V_p = 150$ mV where its value is $r_i = -2.71\ V_p/I_p = -20\ \Omega$.

When voltage V_d across the tunnel diode is given, its current I_{td} can be found from eqs. (3.45). The situation is more difficult when current I_{td} through the tunnel diode is given and we want to find voltage V_d across it, since V_d can not be explicitly expressed from eqs. (3.45) and must be found numerically or graphically. We can also see from Figure 3.19b that when tunnel diode current I_{td} is between I_v and I_p there are three possible solutions for tunnel diode voltage V_p. In one of these solutions tunnel diode voltage V_d is between V_p and V_v: it can be shown that this represents an unstable equilibrium condition. In the other two solutions tunnel diode voltage V_d is below V_p or is above V_v: these represent stable equilibrium conditions.

Example 3.18 Figure 3.20a shows a tunnel diode driven from a current source I_G with a value of $I_G = 0.75\,I_p$. Figure 3.20b shows the I_{td} versus V_d characteristic of the tunnel diode from Figure 3.19b, as well as the $I_{td} = 0.75\,I_p$ straight line imposed by the current source. The three intersections in Figure 3.20b provide $V_d = 2\,V_p$ for the unstable solution, and $V_d = 0.43\,V_p$ and $V_d = 7.05\,V_p$ for the two stable solutions.

The fact that for the current range of $I_v < I_{td} < I_p$ the tunnel diode has two possible stable equilibrium conditions makes it a *two-state* or *bistable* circuit element. Transition from the "*low*" or $V_d < V_p$ state to the "*high*" or $V_d > V_v$ state can be effected by raising I_{td} above I_p. Similarly, transition from the high state to the low state can be effected by lowering I_{td} below I_v. In the absence of capacitance these transitions would be instantaneous.

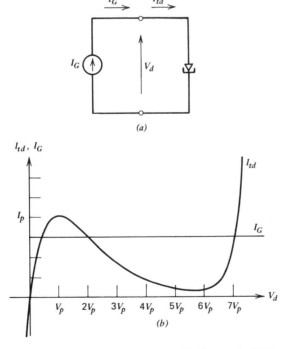

(a)

(b)

Figure 3.20 Tunnel diode driven by a current source. (a) The circuit, (b) Graphical solution for the three values of V_d.

Example 3.19 Figure 3.21a shows a tunnel diode driven by a current source. Source current I_G (see Figure 3.21b) ramps up from $I_G = 0$ to $I_G = 1.75\ I_p$ between times zero and t_1, stays at $I_G = 1.75\ I_p$ between times t_1 and t_2, and ramps back to $I_G = 0$ between times t_2 and t_3. The resulting tunnel diode voltage V_d is shown in Figure 3.21c. It follows the dc current versus voltage characteristics as I_{td} is raised from below I_v to I_p, it jumps instantaneously from V_p to V_f when I_{td} reaches I_p, and it again follows the dc characteristics as I_{td} is further raised. Tunnel diode voltage V_d also follows the dc characteristics as I_{td} is lowered from above I_p to I_v, it jumps instantaneously from V_v to $\approx 0.024\ V_p$ when I_{td} drops to I_v, and it follows the dc characteristics as I_{td} is further lowered.

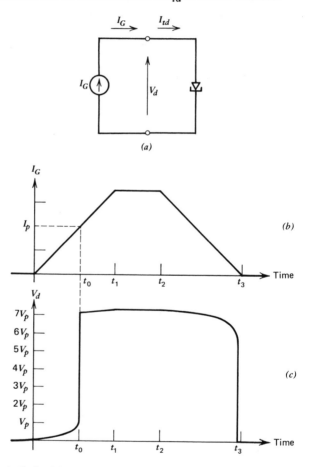

Figure 3.21 Tunnel diode driven by current ramps. (a) The circuit; (b) input current signal; (c) tunnel diode voltage as function of time.

Note that the transition from the low state to the high state takes place at $t = t_0$ when $I_{td} = I_p$, but the transition from the high state to the low state near $t = t_3$ when $I_{td} = I_v$. Thus there is a backlash or *hysteresis* of $I_p - I_v$, and the current swing of I_{td} has to overlap this hysteresis at both ends in order to effect transitions between the two states.

3.3.2 Transients

The tunnel diode circuit and the transients that were shown in Figure 3.21 assumed that stored charges and capacitances were negligible. In reality, as was also the case for junction diodes, stored charges and capacitances originating from several sources are present. In what follows forward transients of tunnel diodes are examined. It can be demonstrated that for such transients the capacitances can be reasonably approximated by a single constant capacitance as shown in Figure 3.22.

As was the case for the similar junction diode transient, the transient in Figure 3.22 can not be found analytically, and a digital computer is used. The flowchart of the computer program is shown in Figure 3.23–it is similar to Figure 3.17, but with a different input current and with eqs. (3.45) used instead of eq. (3.1a).

The resulting transients are illustrated in Figure 3.24 for $I_p t_0/CV_p = 1$ and 10. As expected, the transients become slower as t_0 is increased while all other parameters are held constant. The transients are also slower when capacitance C is increased while all other parameters are held constant, as illustrated in the example that follows.

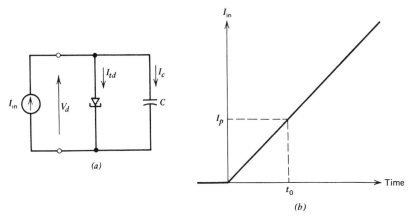

(a)

(b)

Figure 3.22 Simple tunnel diode model driven by current source I_{in}. (a) The circuit; (b) I_{in} as function of time.

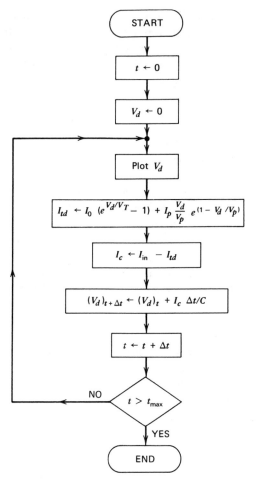

Figure 3.23 Flowchart of the computer program for the transient computation in the circuit of Figure 3.22.

Example 3.20 A germanium tunnel diode is characterized by $V_T = 25$ mV, $V_p = 3V_T = 75$ mV, $I_p = 7.5$ mA, and $I_0 = 5 \times 10^{-10} I_p = 3.75$ pA. The input current ramp I_{in} is characterized by $t_0 = 1$ nsec. Find the time when V_d reaches $4V_p$ if (a) $C = 10$ pF, and (b) $C = 100$ pF.

(a) When $C = 10$ pF,

$$\frac{CV_p}{I_p} = \frac{10 \text{ pF } 75 \text{ mV}}{7.5 \text{ mA}} = 0.1 \text{ nsec.}$$

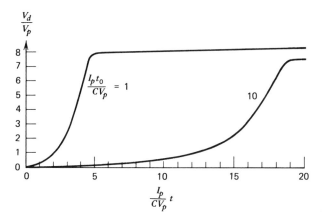

Figure 3.24 Transients in the circuit of Figure 3.22 with $V_p = 3V_T$ and with $I_0 = 5 \times 10^{-10} I_p$.

Also,

$$\frac{I_p t_0}{CV_p} = \frac{7.5 \text{ mA}}{10 \text{ pF}} \frac{1 \text{ nsec}}{75 \text{ mV}} = 10.$$

Reading from Figure 3.24, the $V_d/V_p = 4$ point is reached at $I_p t/CV_p = 16.5$, whence $t = 16.5 \times CV_p/I_p = 16.5 \times 0.1$ nsec $= 1.65$ nsec.

(b) When $C = 100$ pF,

$$\frac{CV_p}{I_p} = \frac{100 \text{ pF}}{7.5 \text{ mA}} = 1 \text{ nsec.}$$

Also,

$$\frac{I_p t_0}{CV_p} = \frac{7.5 \text{ mA}}{100 \text{ pF}} \frac{1 \text{ nsec}}{75 \text{ mV}} = 1.$$

Reading from Figure 3.24, the $V_d/V_p = 4$ point is reached at $I_p t/CV_p = 3.6$, whence $t = 3.6 \times CV_p/I_p = 3.6 \times 1$ nsec $= 3.6$ nsec.

Thus, when $C = 10$ pF, the $V_d = 4 V_p$ point is reached at $t = 1.65$ nsec, and when $C = 100$ pF it is reached at $t = 3.6$ nsec.

PROBLEMS

1. Find the current through the diode described in Example 3.1 when $V_d = -0.1$ V and when $V_d = -0.2$ V.

†2. Derive eqs. (3.2a) and (3.2b).

3. Find r_i and C_d for the diode described in Example 3.5 if it is operated at $I_d = 10$ mA. Check that $r_i C_d$ indeed equals τ.

†4. Derive eqs. (3.4a) and (3.4b).

5. Sketch a rough copy of Figure 3.4 and relabel the axes according to Example 3.6. Find V_d at $t = 10$ psec for Example 3.6 by use of eq. (3.16) and compare the result with your sketch.

6. Sketch diode current I_d and charge Q_d as function of time in Example 3.6.

7. Find the time in Example 3.8 when $V_d = V_F - V_T$.

8. Find the time in Example 3.10 when $V_d = V_F - V_T$.

9. Replace V_g in Example 3.11 by 2 V and find the resulting V_d and I_d graphically.

10. (a) Prepare a flowchart that is similar to the one in Figure 3.12 but is based on eqs. (3.34a) and (3.35a). (b) Use this flowchart for finding the intersection in Figure 3.11b with an accuracy of $\epsilon = 0.2$ mV in V_d. (c) Discuss why the number of steps required by this flowchart is comparable to the seven steps required by the one in Figure 3.12.

11. Replace V_g in Example 3.15 by 2 V and find the resulting V_d and I_d by use of the spiraling-in method.

12. Prepare a flowchart that is similar to the one in Figure 3.14 but is based on eqs. (3.34a) and (3.35b) and show that the resulting spiral drawn into Figure 3.11a diverges.

13. Continue Table 3.3 until $t = 2$ nsec.

†14. Approximate V_d in Figure 3.18 by $V_d \approx -0.1$ V $+ 0.639$ V $\times e^{-t/1 \text{ nsec}}$ Demonstrate that the approximation is very good for all times $t \geqslant 0$ except near $t = 0$. Find the reason for this behavior.

15. The characteristics of a germanium tunnel diode are given by the solid curve of Figure 3.19b with $V_p = 75$ mV and $I_p = 10$ mA. The tunnel diode is driven by the current source shown in Figure 2.9a with $I_0 = 0.75 \, I_p$ and $R_G = 10 \, V_p/I_p$. Find the three solutions for V_d graphically.

16. Sketch the tunnel diode voltage versus time in the circuit of Figure 3.21a if the input current signal of Figure 3.21b is replaced by one that rises to and stays at $0.75 \, I_p$ instead of $1.75 \, I_p$.

†17. Derive the graph of Figure 3.21c from the graph of Figure 3.21b and from Figure 3.19b.

REFERENCES

1. W. F. Chow, *Principles of Tunnel Diode Circuits*, John Wiley and Sons, New York, 1964.
2. J. F. Gibbons, *Semiconductor Electronics*, McGraw-Hill, New York, 1966.
3. A. S. Grove, *Physics and Technology of Semiconductor Devices*, John Wiley and Sons, New York, 1967.
4. J. G. Linvill, *Models of Transistors and Diodes*, McGraw-Hill, New York, 1963.
5. R. S. Muller and T. I. Kamins, *Device Electronics for Integrated Circuits*, John Wiley and Sons, New York, 1977.
6. J. O. Scanlan, *Analysis and Synthesis of Tunnel Diode Circuits*, John Wiley and Sons, New York, 1966.
7. S. M. Sze, *Physics of Semiconductor Devices*, John Wiley and Sons, New York, 1969.

CHAPTER 4

TRANSISTORS AND TRANSISTOR CIRCUITS

Almost without exception, today's high speed digital circuits are based on transistors. The fastest commercially available digital circuits are the Schottky-diode-clamped transistor-transistor logic (TTL) and the emitter-coupled logic (ECL) circuits, and both of these utilize bipolar transistors. For this reason this chapter is limited to bipolar transistors, basic bipolar transistor circuits, and to descriptions of Schottky-diode-clamped TTL and of ECL circuits.

4.1 BIPOLAR TRANSISTORS

A *bipolar transistor*, also known as *junction transistor*, contains two diode junctions. As a result many of its basic properties are related to those of junction diodes, and the discussion that follows draws heavily on the material presented in Chapter 3. The reader should be aware that this section only summarizes the most basic properties of bipolar transistors. More details may be found in the references listed at the end of the chapter.

4.1.1 Basic Structure

A simplified cross section of a bipolar *NPN transistor* is shown in Figure 4.1. Terminals E, B, and C are accessed by way of metallic contacts: terminal E is the *emitter*, terminal B is the *base*, and terminal C is the *collector*. Two diode

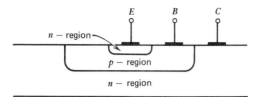

Figure 4.1 Simplified cross-section of an NPN transistor.

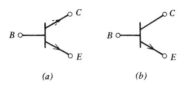

Figure 4.2 Two symbols of an NPN transistor. (*a*) Symbol showing both diodes in the transistor; (*b*) standard symbol.

junctions are formed: one between the base and the emitter, and the other one between the base and the collector.

Two symbols of an NPN transistor are shown in Figure 4.2. The symbol of Figure 4.2*a* shows both diode junctions in the transistor; Figure 4.2*b* shows the commonly used standard symbol.

Figure 4.3 shows the voltage and current polarity conventions that are used. Although they are somewhat arbitrary, they result in voltages and currents that are positive in the majority of the circuits described in this chapter.

When the *n*-regions are replaced by *p*-regions and the *p*-region by an *n*-region, the transistor shown in Figure 4.1 becomes a *PNP transistor*. The operation of a PNP transistor is identical to that of an NPN transistor, but with all voltage and current polarities inverted. In what follows all circuits are illustrated using NPN transistors, even though they are often implemented using PNP transistors.

4.1.2 Basic DC Characteristics

In a bipolar transistor, analogously to eq. (3.1*a*), the dc part of emitter current I_E, $I_{E_{DC}}$, as a function of the dc part of the base-emitter voltage V_{BE}, $V_{BE_{DC}}$, is given by

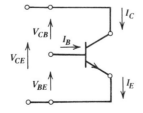

Figure 4.3 Voltage and current polarity conventions used in this chapter.

$$I_{E_{DC}} = I_{ES} \left(e^{V_{BE_{DC}}/V_T} - 1\right), \tag{4.1a}$$

where I_{ES} is the saturation current of the base-emitter diode junction. From eq. (4.1a) it also follows that

$$V_{BE_{DC}} = V_T \ln \left(1 + \frac{I_{E_{DC}}}{I_{ES}}\right). \tag{4.1b}$$

Also, similar to the diode incremental resistance of eq. (3.2), we can define an *incremental emitter resistance* r_e as the slope of the $V_{BE_{DC}}$ versus $I_{E_{DC}}$ curve. Using the property of the exponential function that $d(e^x)/dx = e^x$, we obtain from eq. (4.1a)

$$r_e = \frac{dV_{BE_{DC}}}{dI_{E_{DC}}} = \frac{1}{dI_{E_{DC}}/dV_{BE_{DC}}} = \frac{V_T}{I_{ES}} e^{-V_{BE_{DC}}/V_T} = \frac{V_T}{I_{E_{DC}} + I_{ES}}. \tag{4.2}$$

Example 4.1 In an NPN transistor $I_{ES} = 1$ pA. Find the dc emitter current $I_{E_{DC}}$ and the incremental emitter resistance r_e if the transistor is operated at room temperature with $V_{BE_{DC}} = 0.6$ V.
From eq. (4.1) we get

$$I_{E_{DC}} = I_{ES}(e^{V_{BE_{DC}}/V_T} - 1) = 1 \text{ pA}(e^{0.6 \text{ V}/0.025 \text{ V}} - 1) = 26.5 \text{ mA}.$$

Also, from eq. (4.2) we get

$$r_e = \frac{V_T}{I_{E_{DC}} + I_{ES}} = \frac{25 \text{ mV}}{26.5 \text{ mA} + 1 \text{ pA}} = 0.94 \text{ } \Omega.$$

4.1.3 The Forward Active Operating Region

Consider the case when both V_{BE} and V_{CB} are positive in Figure 4.3. In this case the base-emitter junction is forward biased, and the base-collector junction is reverse biased. Under these conditions the transistor is in its *forward active region* of operation. This region finds widespread use in many circuits.

It is customary to define a *current gain* h_{FE}, also often denoted by β, as

$$h_{FE} = \beta = \frac{I_{C_{DC}}}{I_{B_{DC}}}, \tag{4.3}$$

where $I_{C_{DC}}$ is the dc part of I_C, and $I_{B_{DC}}$ is the dc part of I_B. Another transistor parameter in use is h_{FB}, or α, defined as

$$h_{FB} = \alpha = \frac{I_{C_{DC}}}{I_{E_{DC}}}. \tag{4.4}$$

From eqs. (4.3) and (4.4) it also follows that

$$h_{FE} = \frac{h_{FB}}{1 - h_{FB}} \qquad (4.5a)$$

and that

$$h_{FB} = \frac{h_{FE}}{1 + h_{FE}}, \qquad (4.5b)$$

since $I_{EDC} = I_{CDC} + I_{BDC}$.

In a well-designed transistor that operates in the forward active region, a large part of the dc emitter current originates from the collector and only a small part from the base. Thus $h_{FB} = \alpha$ is near 1, and $h_{FE} = \beta \gg 1$. Typical values of h_{FE} range from 5 to 1000 in high speed transistors.

Example 4.2 The current gain of an NPN transistor is $h_{FE} = 9$. Find the value of α. Also find I_{CDC} and I_{BDC} if $I_{EDC} = 10$ mA.

According to eq. (4.5b),

$$\alpha = h_{FB} = \frac{h_{FE}}{1 + h_{FE}} = \frac{9}{1 + 9} = 0.9.$$

Also, by use of eq. (4.4),

$$I_{CDC} = h_{FB} I_{EDC} = 0.9 \times 10 \text{ mA} = 9 \text{ mA}.$$

Finally, since $I_{EDC} = I_{CDC} + I_{BDC}$,

$$I_{BDC} = I_{EDC} - I_{CDC} = 10 \text{ mA} - 9 \text{ mA} = 1 \text{ mA}.$$

A portion of the collector current originates from the diode current of the reverse-biased base-collector diode. This current, I_{CB}, can be observed separately when the emitter is open-circuited. With the polarity conventions of Figure 4.3 it is given by

$$I_{CB} = -I_{CS}(e^{-V_{CB}/V_T} - 1), \qquad (4.6)$$

where I_{CS} is the saturation current of the base-collector diode. Note that when $V_{CB} \gg V_T$, then $|I_{CB}|$ becomes approximately equal to I_{CS}. Such a small current is sometimes significant in circuits that use transistors operating at low currents; it can, however, usually be neglected in high speed circuits, and is so done in what follows. We also ignore variations of h_{FE} as function of V_{CB} and I_C.

Figure 4.4 outlines the dc collector characteristics of an NPN transistor in the forward active region, based on eqs. (4.3) through (4.5) with $h_{FE} = 10$. Figure 4.5 shows a simple dc model with I_{EDC} as function of V_{BEDC} given by eq. (4.1) and with h_{FB} related to h_{FE} by eqs. (4.3) through (4.5); also, V_{CEDC} is the dc part of V_{CE}.

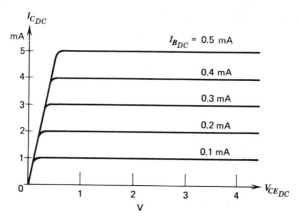

Figure 4.4 Collector characteristics of an NPN transistor based on eqs. (4.3) through (4.5) with $h_{FE} = 10$.

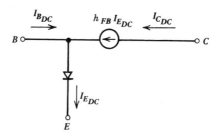

Figure 4.5 Simple dc model of an NPN transistor.

4.1.3.1 Small-Signal Parameters

When the magnitudes of the variations in the currents of Figure 4.5 are much smaller than the magnitudes of the currents themselves, it is often useful to decompose the currents into dc and variable parts as

$$I_E = I_{E_{DC}} + i_e, \tag{4.7a}$$

$$I_B = I_{B_{DC}} + i_b, \tag{4.7b}$$

and

$$I_C = I_{C_{DC}} + i_c. \tag{4.7c}$$

In eqs. (4.7) I_E is the total emitter current, $I_{E_{DC}}$ its dc part, and i_e its varying part; I_B is the total base current, $I_{B_{DC}}$ its dc part, and i_b its varying part; and I_C is

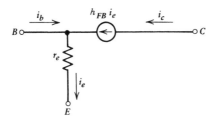

Figure 4.6 Simple small-signal dc model of an NPN transistor using small-signal parameters i_e, i_b, i_c, and h_{FB}.

the total collector current, $I_{C_{DC}}$ its dc part, and i_c its varying part (other symbols are also in use). If we want to focus our attention on the varying parts of the currents, the *small-signal model* shown in Figure 4.6 is often useful.

Note that we used h_{FB} for a small-signal parameter in Figure 4.6. This is justified by the fact that in present-day transistors the dc h_{FB} is very close to the small-signal h_{fb}; also, the dc h_{FE} is very close to the small-signal h_{fe}. In what follows, we distinguish between dc and small-signal parameters only when they are significantly different.

4.1.4 Additional DC Properties

Two additional dc parameters are often important in high speed circuits. One of these is the ohmic resistance r_b in series with the base. Although the resulting dc effects are rarely significant, a nonzero r_b may considerably deteriorate high speed performance.

Another additional dc parameter is the finite output resistance r_{ce} between the collector and the emitter due to the Early effect. As a result eq. (4.3) is modified as

$$I_{C_{DC}} = h_{FE} I_{B_{DC}} \left(1 + \frac{V_{CE_{DC}}}{V_A} \right), \tag{4.8}$$

where V_A is the Early voltage that may be as low as a few volts in high speed transistors.

Example 4.3 A high speed NPN transistor is characterized by $I_{ES} = 1$ pA, $h_{FE} = 10$, $r_b = 1$ kΩ, and $V_A = 4$ V. Sketch the $I_{B_{DC}}$ versus $V_{BE_{DC}}$ characteristic. Also sketch the $I_{C_{DC}}$ versus $V_{CE_{DC}}$ characteristics with $I_{B_{DC}}$ as parameter.

The total base-emitter voltage $V_{BE_{DC}}$ is the sum of $V_{3E_{DC}}$ given by eq. (4.1b) and $r_b I_{B_{DC}}$:

$$V_{BE_{DC}} = V_T \ln \left(1 + \frac{I_{E_{DC}}}{I_{ES}}\right) + r_b I_{B_{DC}}$$

$$= V_T \ln \left(1 + \frac{(1 + h_{FE})I_{B_{DC}}}{I_{ES}}\right) + r_b I_{B_{DC}}.$$

This is sketched in Figure 4.7a. The collector characteristics are given by eq. (4.8) and are shown in Figure 4.7b.

Note in Figure 4.7a that for large base-emitter voltages the base characteristic is dominated by r_b. Also, we can see in Figure 4.7b that the finite V_A makes the lines converge at the $I_{C_{DC}} = 0$, $V_{CE_{DC}} = -V_A$ point. This is in aggrement with eq. (4.8) where $V_{CE_{DC}} = -V_A$ results in $I_{C_{DC}} = 0$.

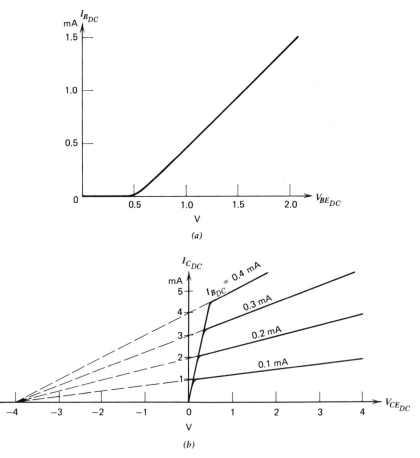

Figure 4.7 Transistor characteristics in Example 4.3. (a) $I_{B_{DC}}$ versus $V_{BE_{DC}}$; (b) $I_{C_{DC}}$ versus $V_{CE_{DC}}$ with $I_{B_{DC}}$ as parameter.

4.1.5 Additional Operating Regions

In addition to the forward active region of operation described above, a bipolar transistor has three additional operating regions. One of these is the *cutoff region* where both the base-emitter and base-collector junctions are reverse biased and only very small currents flow in the transistor. This region is encountered in all types of logic circuits using transistors.

A further region of transistor operation is the *saturated region* where both the base-emitter and the base-collector diodes are forward biased and the collector-to-emitter voltage may be as low as 100 mV. This region finds widespread use in switching circuits of moderate speeds, and its use is also penetrating the realm of high speed circuits.

The fourth region of transistor operation is the *reverse active* region where the base-collector diode is forward biased but the base-emitter diode is reverse biased. This region is rarely used in high speed circuits.

We should note that in practice the delineation of operating regions is not always based exactly on the signs of V_{BE} and V_{CB}. In particular, it is customary to include in the forward active region those parts of the saturated region that result in insignificant I_{CB} in eq. (4.6). Such operating conditions are found, among others, in emitter-coupled logic (ECL) and in Schottky-diode clamped transistor-transistor logic (TTL).

4.1.6 Capacitances, Stored Charge, Gain-Bandwidth Product

In addition to dc characteristics, capacitances in a bipolar transistor are also important when the transistor is used in a high speed circuit. In the forward active region of operation, the base-emitter capacitance is often dominated by the diffusion capacitance of eqs. (3.4), although significant contributions are also made by the transition and stray capacitances. The collector-base capacitance is usually dominated by the transition and stray capacitances in both the forward active and in the cutoff regions. Capacitances in the saturated and reverse active regions are not discussed here, since these regions find only limited use in high speed circuits. In what follows here, the base-emitter capacitance and stored charge are discussed in detail.

The *base-emitter incremental capacitance* C_e can be written as the sum of the base-emitter diffusion capacitance C_{de}, the base-emitter transition capacitance C_{te}, and the base-emitter stray capacitance C_{se}:

$$C_e = C_{de} + C_{te} + C_{se}. \tag{4.9}$$

Based on eqs. (3.4), C_{de} can be written as

$$C_{de} = \frac{\tau_{de} I_{ES}}{V_T} e^{V_{BEDC}/V_T} \tag{4.10a}$$

or as

$$C_{de} = \frac{\tau_{de}}{V_T}(I_{E_{DC}} + I_{ES}). \qquad (4.10b)$$

Timeconstant τ_{de} is characteristic of the transistor and is typically in the range of 10 psec to 1 nsec in high speed transistors. In the simplest case when both C_{te} and C_{se} are negligible, the stored base-emitter charge Q_{BE_d} can be written as

$$Q_{BE_d} = \tau_{de} I_{E_{DC}}. \qquad (4.11)$$

Example 4.4 A high speed transistor is characterized by τ_{de} = 100 psec and I_{ES} = 1 pA. The transistor is operated at a dc emitter current of $I_{E_{DC}}$ = 10 mA. Find the base-emitter diffusion capacitance C_{de} and the stored base-emitter charge Q_{BE_d} at room temperature if $C_{te} \approx 0$ and $C_{se} \approx 0$.
According to eq. (4.10b)

$$C_{de} = \frac{\tau_{de}}{V_T}(I_{E_{DC}} + I_{ES}) = \frac{100 \text{ psec}}{25 \text{ mV}}(10 \text{ mA} + 1 \text{ pA}) \approx 40 \text{ pF}.$$

Also, according to eq. (4.11),

$$Q_{BE_d} = \tau_{de} I_{E_{DC}} = 100 \text{ psec } 10 \text{ mA} = 1 \text{ pC (picocoulomb)}.$$

In general, capacitances C_{te} and C_{se} are not negligible. However, in what follows we approximate C_{te} by a constant, voltage-independent capacitance and include it in C_{se}; thus $C_{te} = 0$ is used from now on. With the foregoing, the total stored charge becomes

$$Q_{BE} = Q_{BE_d} + Q_{BE_s}, \qquad (4.12)$$

where Q_{BE_d} is given by eq. (4.11) and

$$Q_{BE_s} = C_{se} V_{BE}. \qquad (4.13)$$

Hence, by combination of eqs. (4.11), (4.12), and (4.13),

$$Q_{BE} = \tau_{de} I_{E_{DC}} + C_{se} V_{BE}. \qquad (4.14)$$

The contribution to the charge by a nonzero stray capacitance may be significant even when C_{se} is small compared to C_{de}.

Example 4.5 A high speed transistor is characterized by τ_{de} = 100 psec, C_{se} = 0.5 pF, I_{ES} = 1 pA, and V_T = 25 mV. The transistor is operated at a dc emitter current of $I_{E_{DC}}$ = 10 mA. By using eq. (4.1b),

$$V_{BE} = V_T \ln \left(1 + \frac{I_{EDC}}{I_{ES}}\right) = 25 \text{ mV} \ln \left(1 + \frac{10 \text{ mA}}{1 \text{ pA}}\right) = 0.575 \text{ V}.$$

Hence, by use of eq. (4.13),

$$Q_{BE_s} = C_{se} V_{BE} = 0.5 \text{ pF} \times 0.575 \text{ V} = 0.29 \text{ pC}.$$

Also, from Example 4.4 or from eq. (4.11), we have a $Q_{BE_d} = 1$ pC; further, $C_{de} = 40$ pF. Thus $Q_{BE_s}/Q_{BE_d} = 0.29 \text{ pC}/1 \text{ pC} = 29\%$, even though $C_{se}/C_{de} = 0.5 \text{ pF}/40 \text{ pF} = 1.25\%$.

Two additional frequently used transistor parameters are the *characteristic time constant* τ and the *gain-bandwidth product* f_T. The characteristic time-constant is *defined* as

$$\tau \equiv r_e C_e. \tag{4.15}$$

By use of eqs. (4.2), (4.9), and (4.10b), and with $C_{te} = 0$, eq. (4.15) becomes

$$\tau = \tau_{de} + \frac{C_{se} V_T}{I_{EDC} + I_{ES}}. \tag{4.16}$$

The gain-bandwidth product is *defined* as

$$f_T \equiv \frac{1}{2\pi\tau}. \tag{4.17}$$

By use of eq. (4.16), eq. (4.17) becomes

$$f_T = \frac{1}{2\pi\tau_{de}} \frac{1}{1 + (C_{se} V_T/\tau_{de})/(I_{EDC} + I_{ES})}. \tag{4.18a}$$

Note that if $I_{EDC} \gg I_{ES}$, as is the usual case in the forward active operating region, then

$$f_T \approx \frac{1}{2\pi\tau_{de}} \frac{1}{1 + C_{se} V_T/(\tau_{de} I_{EDC})}. \tag{4.18b}$$

Also, by use of eq. (4.10b), eq. (4.18a) can be written as

$$f_T \approx \frac{1}{2\pi\tau_{de}} \frac{1}{1 + C_{se}/C_{de}}. \tag{4.19}$$

According to eqs. (4.18) f_T becomes $1/(2\pi\tau_{de})$ for large values of I_{EDC}, but it gets smaller as I_{EDC} is decreased. This can be also seen in Figure 4.8 where $2\pi\tau_{de}f_T$ is plotted as a function of $\tau_{de}I_{EDC}/(C_{se} V_T)$ according to eq. (4.18b). We can see that $2\pi\tau_{de}f_T = 1$ when $\tau_{de}I_{EDC}/(C_{se} V_T) \gg 1$, and it becomes $\approx \tau_{de}I_{EDC}/(C_{se} V_T)$ when $\tau_{de}I_{EDC}/(C_{se} V_T) \ll 1$.

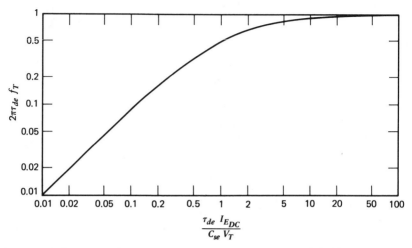

Figure 4.8 Plot of $2\pi\tau_{de}f_T$ as a function of $\tau_{de}I_{E_{DC}}/(C_{se}V_T)$.

A plot of gain-bandwidth product f_T as a function of $I_{E_{DC}}$ is often given in transistor data sheets. Such a plot can provide a basis for finding τ_{de} and C_{se} when V_T is known.

Example 4.6 Figure 4.9 shows f_T as a function of $I_{E_{DC}}$ for a high speed transistor. The decrease of f_T at very high currents is due to current limiting in the transistor and is ignored here. Also, we assume that the decrease

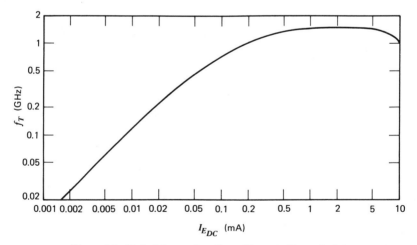

Figure 4.9 Plot of f_T as a function of $I_{E_{DC}}$ in Example 4.6.

of f_T at low currents is due entirely to nonzero C_{se}. We now find τ_{de} and C_{se} for $V_T = 25$ mV.

We can see that f_T becomes ≈ 1.6 GHz for large values of $I_{E_{DC}}$. Thus, $2\pi\tau_{de} \times 1.6$ GHz $= 1$, whence $\tau_{de} \approx 100$ psec. Next we must "match" Figure 4.9 with Figure 4.8 horizontally. This can be done at least two different ways.

(1) We could align the two curves at low currents where $2\pi\tau_{de} f_T \approx \tau_{de} I_{E_{DC}}/(C_{se} V_T)$, that is, $f_T \approx I_{E_{DC}}/(2\pi C_{se} V_T)$. However our data does not extend to low enough currents to permit a good match. Also, at very low currents f_T is often influenced by effects other than C_{se}.

(2) A better way is to match at the $2\pi\tau_{de} f_T = 0.5$ points, which occur at $\tau_{de} I_{E_{DC}}/(C_{se} V_T) = 1$. Since the maximum value of f_T is 1.6 GHz, $2\pi\tau_{de} f_T = 0.5$ when $f_T = 0.8$ GHz, which takes place at $I_{E_{DC}} = 0.125$ mA in Figure 4.9. Thus, since $\tau_{de} I_{E_{DC}}/(C_{se} V_T) = 1$ at this point, $C_{se} = \tau_{de} I_{E_{DC}}/V_T = 100$ psec $\times$ 0.125 mA/25 mV $= 0.5$ pF.

4.1.7 Transistor Models

Two simple dc models for NPN transistors were shown in Figures 4.5 and 4.6. In what follows here, we describe four models that include all effects discussed in Sections 4.1.1 through 4.1.6.

Figure 4.10*a* shows a general *hybrid-π* (or *hybrid-pi*) model. It is also known as a *large-signal* model, since its use is not restricted to small-signal operation. In contrast, the hybrid-π transistor model of Figure 4.10*b* is a small-signal model that is usable only for small-signal operation.

Two additional transistor models are shown in Figure 4.11. It can be shown that, as observed at terminals E, B, and C, the model of Figure 4.11*a* is equivalent to that of Figure 4.10*a*, and the model of Figure 4.11*b* is equivalent to that of Figure 4.10*b*.

The models of Figures 4.10 and 4.11 are often enlarged by additional circuit elements in order to provide more accurate representations of today's high speed transistors. One of these elements is an ohmic collector resistance in series with collector terminal C; this resistance is usually dominated by the resistance of the collector *n*-region in the transistor. Another additional circuit element is an ohmic emitter resistance in series with emitter terminal E; this resistance is usually dominated by the resistance that appears between the emitter *n*-region and the metallic emitter contact. In a packaged transistor additional significant capacitances may be present between the connecting wires; in a transistor that is part of an integrated circuit, the capacitance between the collector and the semiconductor substrate is often significant.

The models just discussed are quite complex, even though they are restricted to the forward active region of an NPN transistor. However, these models can be simplified in many applications. As a simple example, all capacitances can be omitted if we are interested only in dc operation. Models where some, but not all, capacitances are omitted are introduced later in this chapter.

The choice between the models of Figure 4.10 versus those of Figure 4.11 is a matter of convenience. As for the choice between a general (large-signal) and a small-signal model, the latter permits the utilization of the simpler linear analysis techniques discussed in Chapter 2. Thus, a small-signal model should be used whenever this is made permissible by "small" signals in the circuit.

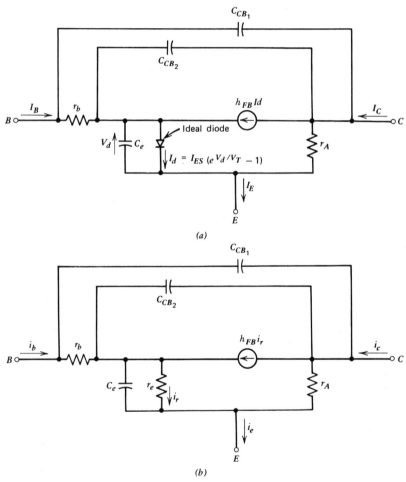

Figure 4.10 Hybrid-π transistor models. (a) General (large-signal) model; (b) small-signal model.

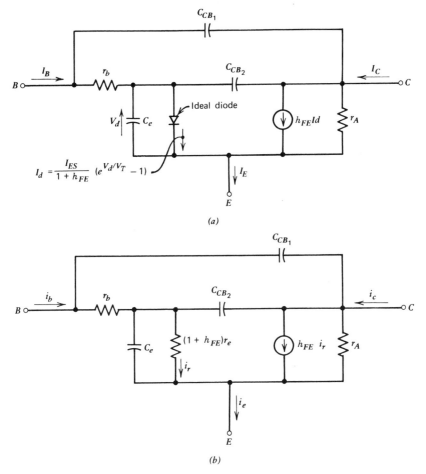

Figure 4.11 Alternate transistor models. (*a*) General (large-signal) model; (*b*) small-signal model.

4.2 BASIC TRANSISTOR CIRCUITS

This section describes three basic bipolar transistor circuits: the grounded base stage, the grounded emitter stage, and the emitter follower (grounded collector stage). The discussion is limited to the forward active region, and it is assumed that power supply voltages are chosen to provide such operation. Emphasis is placed on circuit characteristics that are important for use in high speed digital logic such as Schottky-diode-clamped TTL and emitter-coupled logic (ECL).

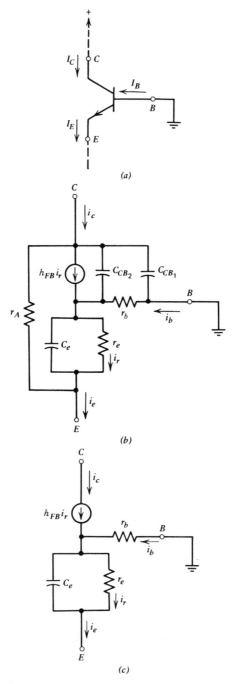

Figure 4.12 Grounded base stage. (*a*) The circuit; (*b*) equivalent circuit using the small-signal model of Figure 4.10*b* with $r_A = \infty$; (*c*) simplified equivalent circuit.

4.2.1 The Grounded Base Stage

Figure 4.12a shows the circuit of a grounded base stage, Figure 4.12b an equivalent circuit utilizing the small-signal model of Figure 4.10b. We simplify Figure 4.12 by omitting C_{CB_1}, C_{CB_2}, and r_A, leading to the simplified equivalent circuit shown in Figure 4.12c. (A nonzero C_{CB_1} is included in Problem 7).

4.2.1.1 Small-Signal Response

We now look at the small-signal component of the collector current in the grounded base stage described by Figure 4.12c. We assume that the drive is a current source as shown in Figure 4.13, where currents I_E, I_B, and I_C are decomposed into dc and into variable small-signal parts in accordance with Eqs. (4.7).

First we consider the case when the small-signal part of current source

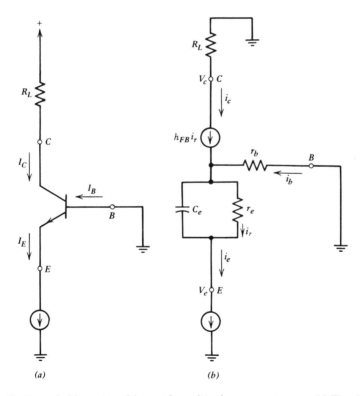

(a)   (b)

Figure 4.13 Grounded base stage driven at its emitter by a current source. (a) The circuit; (b) small-signal equivalent circuit.

I_E, i_e, is a step current:

$$i_e = i_{e_0} u(t). \tag{4.20}$$

By considering the transient in Figure 2.16b we can write for the circuit of Figure 4.13:

$$i_r = i_{e_0} (1 - e^{-t/\tau}), \tag{4.21a}$$

$$i_c = h_{FB} i_r = h_{FB} i_{e_0} (1 - e^{-t/\tau}), \tag{4.21b}$$

where, in accordance with eq. (4.15),

$$\tau = r_e C_e. \tag{4.21c}$$

Next we consider the case when the small-signal part of current source I_E, i_e, is a sinewave:

$$i_e = i_{e_0} \sin(2\pi ft). \tag{4.22}$$

In this case, based on Section 2.6.4, we can write

$$|i_c| = \frac{h_{FB} |i_{e_0}|}{\sqrt{1 + (f/f_T)^2}}, \tag{4.23a}$$

$$\underline{/i_c} = -\arctan(f/f_T), \tag{4.23b}$$

where f_T is given by eq. (4.17).

Example 4.7 In the circuit of Figure 4.13 the transistor is characterized by $h_{FB} = 0.9$ and $\tau = 100$ psec. Current source $I_E = I_{E_{DC}} + i_e$; $I_{E_{DC}} = 10$ mA. Characterize I_C if (a) $i_e = 0.1$ mA $u(t)$, (b) $i_e = 0.1$ mA $\sin(2\pi 800$ MHz $t)$.

The value of f_T is $f_T = 1/(2\pi\tau) = 1/(2\pi 100$ psec$) \approx 1.6$ GHz. Also $I_C = I_{CDC} + i_c$ where $I_{CDC} = h_{FB} I_{EDC} = 0.9 \times 10$ mA $= 9$ mA. Thus, for (a) $i_c = 0.09$ mA$(1 - e^{-t/100 \text{ psec}})$, and for (b) $|i_c| = 0.09$ mA/$\sqrt{1 + (800/1600)^2} = 0.08$ mA and $\underline{/i_c} = -\arctan(800/1600) = -26.6°$.

The results obtained above for the small-signal response can also be used for large signals when τ of eq. (4.21c) is independent of the emitter current, provided that the transistor remains in its forward active region of operation. However, τ is given by eq. (4.16) at a given $I_{E_{DC}}$, and it is assumed that I_E remains close to $I_{E_{DC}}$. When the change in I_E, i_e, is not small compared to $I_{E_{DC}}$, eq. (4.16) is not applicable. In such cases we can often use an "average" value of τ and obtain reasonable approximations—this is discussed later.

4.2.1.2 Small-Signal Input Impedance

When the emitter of a grounded base stage is driven by a source other than a current source, the impedance seen looking into the emitter is of interest. This

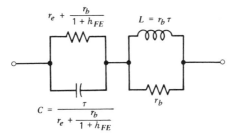

Figure 4.14 Small-signal input impedance of a grounded base stage with $C_{CB_2} = 0$.

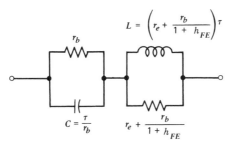

Figure 4.15 Series compensating network for the input of a grounded base stage with the input impedance of Figure 4.14.

input impedance is given without proof in Figure 4.14. We can see that when $r_b = 0$, the input impedance reduces to a resistance r_e in parallel with a capacitance of τ/r_e. However, the impedance becomes more complicated when $r_b \neq 0$.

Often it is desirable to have a purely resistive input impedance to a grounded base stage. This can be accomplished by connecting the network shown in Figure 4.15 in series with the emitter. The result is a pure resistance with a magnitude of $r_e + r_b + r_b/(h_{FE} + 1)$.

4.2.2 The Grounded Emitter Stage

Figure 4.16a shows the circuit of a grounded emitter stage, Figure 4.16b an equivalent circuit utilizing the small-signal model of Figure 4.11b with $C_{CB_1} = 0$, $C_{CB_2} = 0$, and $r_A = \infty$. In the following sections we discuss the small-signal and large-signal responses. We do not discuss input impedances, since these can be found directly from Figure 4.11 when $C_{CB_1} = C_{CB_2} = 0$.

4.2.2.1 Small-Signal Response

We now look at the small-signal part of the collector current in the grounded emitter stage described by Figure 4.16b, when driven by a current source as

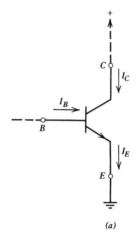

(a)

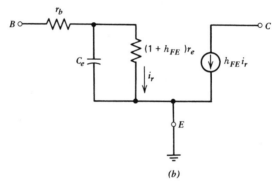

(b)

Figure 4.16 Grounded emitter stage. (a) The circuit; (b) equivalent circuit using the small-signal model of Figure 4.11b with $C_{CB_1} = 0$ and $C_{CB_2} = 0$.

shown in Figure 4.17. In what follows, currents I_E, I_B, and I_C are decomposed into dc and into variable small-signal parts in accordance with eqs. (4.7).

We consider here only the case when the small-signal part of current source I_B, i_b, is a step current:

$$i_b = i_{b_0} u(t); \qquad (4.24)$$

the response to a sinewave input is given in Problem 11.

By considering the transient in Figure 2.16b we can write for the circuit of Figure 4.17b:

$$i_c = i_{b_0} h_{FE} \{1 - e^{-t/[(h_{FE} + 1)\tau]}\}, \qquad (4.25a)$$

where, in accordance with eq. (4.15),

$$\tau = r_e C_e. \qquad (4.25b)$$

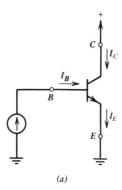

(a)

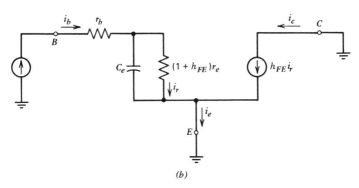

(b)

Figure 4.17 Grounded emitter stage driven at its base by a current source. (a) The circuit; (b) small-signal equivalent circuit.

Example 4.8 In the circuit of Figure 4.17 the transistor is characterized by $h_{FE} = 100$ and $\tau = 200$ psec. Current source $I_B = I_{B_{DC}} + i_b$; $I_{B_{DC}} = 0.1$ mA, $i_b = 2\,\mu A\,u(t)$.

Thus, $I_{C_{DC}} = 0.1$ mA $\times 100 = 10$ mA. Also, by use of eq. (4.25a), $i_c = 2\,\mu A\,100\,\{1 - e^{-t/[(100+1)\,200\text{ psec}]}\} = 0.2$ mA $\{1 - e^{-t/20.2\text{ nsec}}\}$.

We can also approximate eq. (4.25a) by use of the expansion of eq. (2.25) and retain only the first two terms. Thus, $e^x \approx 1 + x$ with $x = -t/[(h_{FE} + 1)\tau]$. This results in

$$i_c \approx i_{b_0} \frac{h_{FE}}{h_{FE} + 1} \frac{t}{\tau} = i_{b_0} h_{FB} \frac{t}{\tau}, \qquad (4.26a)$$

which is valid when

$$t \ll (h_{FE} + 1)\tau. \qquad (4.26b)$$

Example 4.9 Find i_c at $t = 2$ nsec in the circuit of Example 4.8. Use the approximate eqs. (4.26) and compare the result with the exact i_c.

First we check the condition of eq. (4.26b): $(h_{FE} + 1)\tau = (100 + 1)200$ psec $= 20.2$ nsec, significantly greater than $t = 2$ nsec. Thus, by use of eq. (4.26a),

$$i_c \approx i_{b_0} \frac{h_{FE}}{h_{FE} + 1} \frac{t}{\tau} = 2\ \mu A \frac{100}{100 + 1} \frac{2\ \text{nsec}}{200\ \text{psec}} = 19.8\ \mu A.$$

The exact i_c, by use of i_c from Example 4.8,

$$i_c = 2\ \mu A \times 100 \times \{1 - e^{-2\ \text{nsec}/20.2\ \text{nsec}}\} = 18.85\ \mu A.$$

Thus, in this example, the use of the approximate eqs. (4.26) results in a fractional error of $(19.8\ \mu A - 18.85\ \mu A)/18.85\ \mu A \approx 5\%$.

4.2.2.2 Large-Signal Response

Often, unlike in the preceding Section 4.2.2.1, the variable parts of the currents are not small compared to the dc currents. The results, eqs. (4.25) and (4.26), are still valid when τ is current-independent, that is, when the second term is negligible in eq. (4.14). When this is not the case computer-aided methods, such as described in Section 3.2.2, can be used to find the transient.

In many cases, however, we would like to *approximate* the transient in the form of eqs. (4.25) and (4.26) even when $C_{se} \neq 0$. This can be done by replacing τ in eqs. (4.25) and (4.26) by an *average characteristic timeconstant* τ_{ave}, which is defined as follows:

Assume that a transistor is switched between emitter currents I_{E_1} and I_{E_2}. Thus we also have, following eq. (4.1b),

$$V_{BE_1} = V_T \ln\left(1 + \frac{I_{E_1}}{I_{ES}}\right), \tag{4.27a}$$

$$V_{BE_2} = V_T \ln\left(1 + \frac{I_{E_2}}{I_{ES}}\right). \tag{4.27b}$$

We now define τ_{ave} as

$$\tau_{ave} \equiv \tau_{de} + C_{se} \frac{V_{BE_2} - V_{BE_1}}{I_{E_2} - I_{E_1}}. \tag{4.28}$$

Example 4.10 A transistor is characterized by τ_{de} = 100 psec, C_{se} = 0.5 pF, I_{ES} = 1 pA, and V_T = 25 mV. Find τ_{ave} if the transistor is switched between (a) I_{E_1} = 0 and I_{E_2} = 10 mA, (b) I_{E_1} = 10 μA and I_{E_2} = 10 mA.

(a) According to eqs. (4.27), V_{BE_1} = 0 and V_{BE_2} = 0.575 V. Thus, from eq. (4.28), τ_{ave} = 100 psec + 0.5 pF × 0.575 V/10 mA =129 psec. Hence, in this case, τ_{ave} is greater than τ_{de} by 29%. This is identical to the Q_{BE_s}/Q_{BE_d} = 29% in Example 4.5. Also, a comparison of eq. (4.28) with eq. (4.14) shows that this is true in general as long as V_{BE_1} = 0 *and* I_{E_1} = 0 (or V_{BE_2} = 0 and I_{E_2} = 0).

(b) In this case V_{BE_1} = 0.403 V and V_{BE_2} = 0.575 V. Thus, from eq. (4.28), τ_{ave} = 100 psec + 0.5 pF (0.575 V - 0.403 V)/(10 mA - 10 μA) = 108.6 psec. Thus, in this case, τ_{ave} is greater than τ_{de} by 8.6%.

4.2.3 The Emitter Follower

Figure 4.18a shows the circuit of an emitter follower, which is also known as a grounded collector stage. Figure 4.18b shows an equivalent circuit using the small-signal model of Figure 4.10b with C_{CB_1} = 0, C_{CB_2} = 0, and r_A = ∞. In the following sections we discuss impedances, transient responses, and stability.

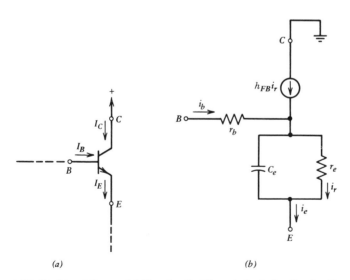

(a)   (b)

Figure 4.18 Emitter follower. (a) The circuit; (b) equivalent circuit using the small-signal model of Figure 4.10b with C_{CB_1} = 0, C_{CB_2} = 0, and r_A = ∞.

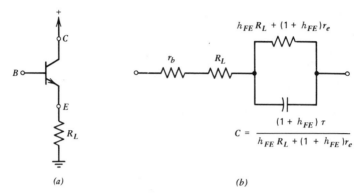

Figure 4.19 Small-signal input impedance of an emitter follower loaded at its output by a resistance R_L. (*a*) The circuit; (*b*) small-signal input impedance.

4.2.3.1 Input Impedance with Resistive Load

Figure 4.19 shows the input impedance seen at base B when the load is a resistance R_L. Since the voltage across a capacitance can not change instantly, for very short times after the start of a transient the input impedance is $r_b + R_L$; hence, the emitter follower does nothing. However, for times $t \gg (h_{FE} + 1)\tau$ after the start of a transient the input impedance becomes $r_b + (h_{FE} + 1)(r_e + R_L)$.

Example 4.11 A transistor is characterized by $h_{FE} = 100$, $\tau = 150$ psec, $r_b = 100\ \Omega$, and $r_e = 2.5\ \Omega$. It is operated in the circuit of Figure 4.19*a* with a load resistance of $R_L = 50\ \Omega$.

The input impedance is shown in Figure 4.20. For very short times after the start of a transient it becomes $100\ \Omega + 50\ \Omega = 150\ \Omega$; for times much longer than $(100 + 1)\,0.15$ nsec = 15.15 nsec it becomes $100\ \Omega + 50$ $\Omega + 5252.5\ \Omega = 5402.5\ \Omega$. Note that this is much greater than the load resistance of $R_L = 50\ \Omega$.

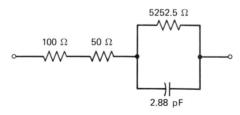

Figure 4.20 Input impedance in Example 4.11.

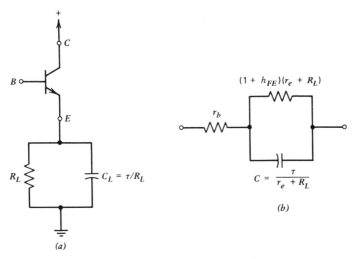

Figure 4.21 Emitter follower loaded by a resistance R_L in parallel with a capacitance $C_L =$ τ/R_L. (a) The circuit; (b) small-signal input impedance.

4.2.3.2 Input Impedance with Parallel R-C Load

The input impedance of an emitter follower becomes quite complicated when its output is loaded by a resistance R_L in parallel with a capacitance C_L. An exception to this occurs when load capacitance C_L has a value of τ/R_L, as shown in Figure 4.21a. In this special case the small-signal input impedance becomes the simple circuit shown in Figure 4.21b. Note that this input impedance is similar to the input impedance of a grounded emitter stage, but with r_e replaced by $r_e + R_L$. Also, as expected, for times $t \gg (h_{FE} + 1)\tau$ after the start of a transient this input impedance becomes identical to the input impedance of $r_b + (h_{FE} + 1)(r_e + R_L)$ of Figure 4.19.

4.2.3.3 Output Impedance with Resistive Source

When the input of an emitter follower is driven from a source that has a purely resistive source impedance R_G, the small-signal output impedance seen at the emitter is given by the circuit of Figure 4.14 with r_b replaced by $R_G + r_b$.

Example 4.12 The transistor described in Example 4.11 is used as an emitter follower and is driven from a resistive source with a source resistance of $R_G = 400 \; \Omega$. Thus, for times $t \gg \tau$ after the start of a transient the output impedance becomes $2.5 \; \Omega + (400 \; \Omega + 100 \; \Omega)/(100 + 1) = 7.45$ Ω. Note that this is much less than the source resistance of $R_G = 400 \; \Omega$.

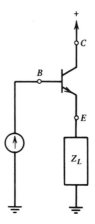

Figure 4.22   Emitter follower driven by a current source.

4.2.3.4 Transient with Current Source Input

Figure 4.22 shows an emitter follower driven at its input by a current source that has a small-signal part of $i_b = i_{b_0} u(t)$, and is loaded at its output by a finite load impedance Z_L. By use of the small-signal model of Figure 4.10b with $C_{CB_1} = 0$, $C_{CB_2} = 0$, and $r_A = \infty$, the small-signal part of the output current can be found as

$$i_e = i_{b_0} \left\{ 1 + h_{FE} \left[1 - e^{-t/[(h_{FE} + 1)\tau]} \right] \right\}. \tag{4.29}$$

Also, for times $t \ll (h_{FE} + 1)\tau$ after the start of a transient we can use eq. (2.25) for approximating $e^x \approx 1 + x$ with $x = -t/[(h_{FE} + 1)\tau]$. This results in

$$i_e \approx i_{b_0} \left(1 + \frac{h_{FE}}{h_{FE} + 1} \frac{t}{\tau} \right) = i_{b_0} \left(1 + h_{FB} \frac{t}{\tau} \right). \tag{4.30}$$

Example 4.13   A transistor is characterized by $h_{FE} = 100$ and $\tau = 200$ psec. It is operated in the circuit of Figure 4.22 with $i_b = 2.5$ mA $u(t)$. Find i_e at $t = 5$ nsec. By use of the exact eq. (4.29) we get $i_e (t = 5 \text{ nsec}) = 57.3$ mA, by use of the approximate eq. (4.30), $i_e (t = 5 \text{ nsec}) = 64.4$ mA. Thus, in this example, the use of the approximate eq. (4.30) results in a fractional error of $(64.4 - 57.3)/57.3 = 0.12 = 12\%$.

Note that eqs. (4.29) and (4.30) are also valid when the magnitude of i_{b_0} is not small compared to $I_{B_{DC}}$, provided that τ can be approximated as constant and that the transistor remains in its forward active region of operation.

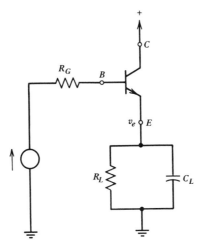

Figure 4.23 Emitter follower driven by a voltage source and loaded by the parallel combination of resistance R_L and capacitance C_L.

4.2.3.5 Transient with Voltage Source Input

Figure 4.23 shows an emitter follower driven at its input by a voltage source in series with a source resistance R_G; the small-signal part of the voltage source is $v_g = v_{g_0} u(t)$. The output of the emitter follower is loaded by a resistance R_L in parallel with a capacitance C_L.

The transient response of v_e is somewhat complicated, except when $C_L = \tau/R_L$. In this special case, the small-signal part of the output voltage can be approximated for $(h_{FE} + 1)(R_L + r_e) \gg R_G + r_b$ as

$$v_e \approx v_{g_0} \frac{R_L}{R_L + r_e} \{1 - e^{-t/[\tau(R_G + r_b)/(R_L + r_e)]}\}. \qquad (4.31)$$

Example 4.14 A transistor is characterized by $h_{FE} = 100$, $\tau = 150$ psec, $r_b = 100\ \Omega$, and $r_e = 2.5\ \Omega$. It is operated in the circuit of Figure 4.23 with $R_G = 400\ \Omega$, $R_L = 50\ \Omega$, and $C_L = 3$ pF.

Thus, $C_L = \tau/R_L$. Also, $(h_{FE} + 1)(R_L + r_e) = 5302.5\ \Omega$, which is much greater than $R_G + r_b = 500\ \Omega$. Hence, eq. (4.31) yields the approximation of $v_e = 0.95\ v_{g_0}(1 - e^{-t/1.43 \text{ nsec}})$. Thus, for times $t \gg 1.43$ nsec after the start of a transient, output voltage v_e becomes $0.95\ v_{g_0}$ —hence the name "emitter follower."

4.2.3.6 Instabilities

An emitter follower may become unstable under certain conditions and oscillate even when no input signal is provided. Figure 4.24 shows an emitter follower that is driven by a voltage source in series with an inductance L—note that the voltage source is identically zero; hence, it is equivalent to a short circuit. The output of the emitter follower is loaded by a capacitance C_L. It can be shown by use of the small-signal model of Figure 4.10b with $C_{CB_1} = 0$, $C_{CB_2} = 0$, and $r_A = \infty$, that the circuit of Figure 4.24 oscillates when

$$L \geqslant r_b \tau (h_{FE} + 1) \frac{h_{FE} + 1 + (C_L r_b / \tau) \left[1 + (r_e / r_b)(h_{FE} + 1) \right]}{h_{FE}(h_{FE} + 1) - (C_L r_b / \tau) \left[1 + (r_e / r_b)(h_{FE} + 1) \right]} \geqslant 0; \tag{4.32a}$$

however, the circuit is stable with any positive L when the denominator in eq. (4.32a) is negative, that is, when

$$\frac{C_L r_b}{\tau} \left[1 + \frac{r_e}{r_b}(h_{FE} + 1) \right] > h_{FE}(h_{FE} + 1). \tag{4.32b}$$

Example 4.15   A transistor is characterized by $h_{FE} = 100$, $\tau = 150$ psec, $r_b = 100$ Ω, and $r_e = 2.5$ Ω, and it is used in the circuit of Figure 4.24. According to eq. (4.32b), the circuit is stable with any positive L when

$$C_L > \frac{\tau}{r_b} \frac{h_{FE}(h_{FE} + 1)}{1 + (r_e / r_b)(h_{FE} + 1)} \approx 4300 \text{ pF}.$$

This is a large load capacitance, and it would result in slow transients. With a load capacitance of, for example, $C_L = 3$ pF the circuit is stable, accord-

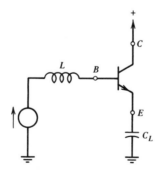

Figure 4.24   Emitter follower driven by an inductive source L and loaded by a capacitance C_L.

ing to eq. (4.32a), only when $L < 16.2$ nH—which is easily exceeded by wiring inductance if adequate care is not exercised.

Equations (4.32) are also applicable when there is a source resistance R_G in series with inductance L. In such a case, r_b in eqs. (4.32) must be replaced by $R_G + r_b$. Note that the addition of R_G permits the use of a larger L in eq. (4.32a) and a smaller C_L in eq. (4.32b). (See also Problem 13)

4.3 SCHOTTKY-DIODE-CLAMPED TTL

Schottky-diode-clamped TTL, also known as Schottky TTL, is the fastest available logic family that is compatible with other transistor-transistor logic (TTL). In the following sections we discuss basic properties, noise margins, dc operation, and propagation delays.

4.3.1 Basic Properties

Figure 4.25 shows the symbol of a *logic inverter*. As customary, the symbol does not show the connections of the ground and the +5 V power supply. The dc transfer characteristics of a Schottky-diode-clamped TTL inverter are shown in Figure 4.26. We can see that there is a sharp transition at an input voltage of about +1.5 V. Also, the output voltage is insensitive to changes in the input voltage when the input voltage is either less than about +1 V or when it is greater than about +2 V.

Specifications on the output voltage are guaranteed for the case when the output is loaded by the inputs of 10 similar circuits as follows: When the input voltage is at or below +0.8 V, the output voltage is at least +2.5 V; when the input voltage is at or above +2 V, the output voltage is at most +0.5 V.

Figure 4.27a shows a symbol for a 2-input NAND logic gate, and Figure 4.27b a *truth table* summarizing its operation. We can see that the output is "LOW" only when both inputs are "HIGH"--hence the name NAND for "NOT AND."

The logic inverter and the 2-input NAND logic gate are only the simplest of the wide variety of logic circuits that are available. The reader is referred to the available (and ever expanding) catalogs for further data.

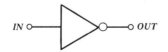

Figure 4.25 Symbol of a logic inverter.

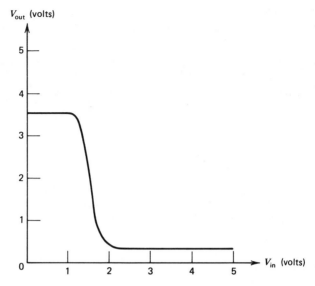

Figure 4.26 Dc transfer characteristics of a Schottky-diode-clamped TTL inverter.

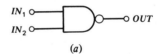

(a)

Inputs		Output
IN_1	IN_2	OUT
⩽0.8 V	⩽0.8 V	⩾2.5 V
⩽0.8 V	⩾2.0 V	⩾2.5 V
⩾2.0 V	⩽0.8 V	⩾2.5 V
⩾2.0 V	⩾2.0 V	⩽0.5 V

(b)

Figure 4.27 Schottky-diode-clamped TTL 2-input NAND gate. (a) Symbol; (b) truth table.

4.3.2 Noise Margins

From the specifications in the preceding section we can conclude that there is some leeway in the voltage levels when the output of a logic gate is connected to the inputs of other logic gates. The maximum possible output voltage in a LOW output state is +0.5 V, while an input can tolerate a maximum of +0.8 V—

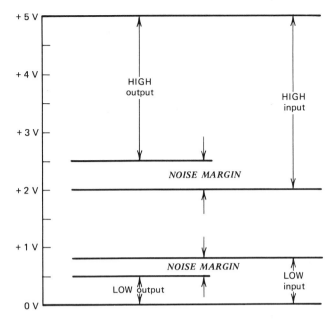

Figure 4.28 Band diagram showing noise margins in Schottky-diode-clamped TTL circuits.

thus, there is a leeway, or *noise margin*, of 0.8 V – 0.5 V = 0.3 V. Similarly, the minimum output voltage in the HIGH output state is +2.5 V, while an input can tolerate a minimum of +2 V—these result in a noise margin of 2.5 V – 2 V = 0.5 V.

The noise margins are provided to accommodate noise and interference picked up along the interconnections between the logic gates. These may originate from voltage drops along ground and power lines and from capacitive or inductive pickup, as well as from reflections that can be present when the circuits are interconnected by a *transmission line*—this is discussed in Chapter 5.

It is often desirable to show the noise margins graphically. One such presentation is the *band diagram*, shown in Figure 4.28 for Schottky-diode-clamped TTL circuits.

4.3.3 DC Operation

We have seen in Figure 4.27*b* that the output of a Schottky-diode clamped TTL NAND gate is LOW when both inputs are HIGH and the output is HIGH otherwise. Here we describe the dc operation of the circuit for the second and the fourth (last) rows of the truth table shown in Figure 4.27*b*.

The dc conditions corresponding to the second row in Figure 4.27*b* are shown in Figure 4.29. Diodes D_1 and D_2 at the inputs limit reflections that may be present when an interconnection is made by a transmission line—this is

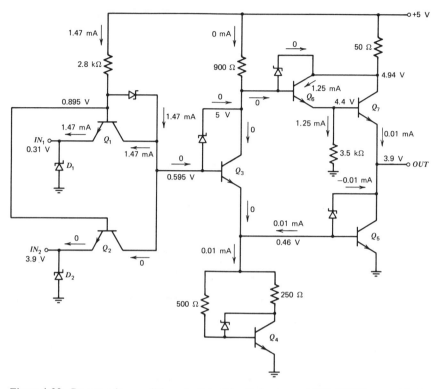

Figure 4.29 Dc operating condition of a Schottky-diode-clamped TTL NAND gate with the output HIGH.

discussed in detail in Chapter 5. The remainder of the Schottky diodes clamp collector-base junctions, limiting the forward junction voltages to about 0.3 V. Such a voltage is too small to result in a significant forward current in the collector-base junction; hence, the transistor can be approximated as being in its forward active region of operation. Also, we assume that all transistors have I_{ES} = 0.1 pA and that they operate at room temperature, whereby V_T = 25 mV.

Figure 4.29 includes another simplification: As a crude approximation, we neglect all base currents—it can be shown that this does not result in a significant loss of accuracy. The dc conditions corresponding to the fourth (last) row of Figure 4.27*b* are shown in Figure 4.30 with the same approximations as were used in Figure 4.29.

The operation of transistor Q_4 is detailed in Figure 4.31. The circuit of Figure 4.31*a* replaces a resistor of ≈4.6 kΩ. Such a resistor would provide the same pulldown current of I_1 ≈ 1.35 mA at V_1 = 0.61 V as the circuit of Figure 4.31*a*. However, according to Figure 4.31*b*, when I_1 is lowered to ≈0.01 mA,

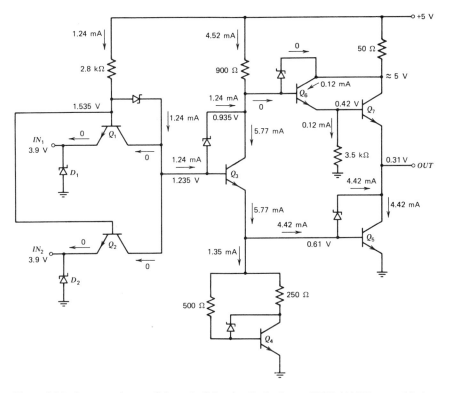

Figure 4.30 Dc operating condition of a Schottky-diode-clamped TTL NAND gate with the output LOW.

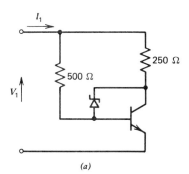

(a)

Figure 4.31 Operation of Q_4 in Figure 4.29 and 4.30. (*a*) The circuit; (*b*) dc characteristics. (See page 126 for Figure 4.31*b*.)

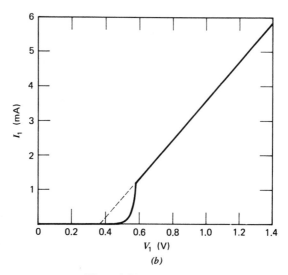

Figure 4.31 (continued)

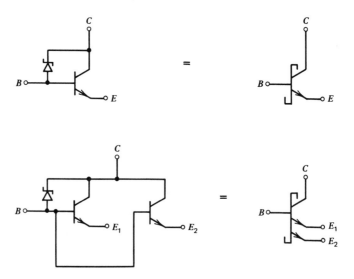

Figure 4.32 Alternative component symbols used in Schottky-diode-clamped TTL circuit diagrams.

the circuit of Figure 4.31a provides a $V_1 \approx 0.46$ V, while a 4.6 kΩ resistor would provide only $\approx$46 mV. Thus, the voltage swing in the circuit of Figure 4.31a (and also between Figures 4.29 and 4.30) is only 0.15 V, while across the 4.6 kΩ resistor it would be slightly over 0.6 V. Thus, the charge that must be applied to the stray capacitances during switching is reduced by a factor of 4 by the use of the circuit of Figure 4.31a.

Finally, we should mention that the symbols shown in Figure 4.32 are often used in circuit diagrams of Schottky-diode-clamped TTL.

4.3.4 Propagation Delays

Propagation delays in Schottky-diode-clamped TTL gates are usually specified with the output loaded by the dc input currents of 10 similar gates, as well as by a load capacitance C_L. Propagation delays are specified at the +1.5 V level as follows.

Propagation delays for a LOW to HIGH output transition t_{PLH} are 3 nsec typical and 4.5 nsec maximum when C_L = 15 pF; when C_L = 50 pF, t_{PLH} = 4.5 nsec typical, and no maximum is given. Propagation delays for a HIGH to LOW output transition t_{PHL} are 3 nsec typical and 5 nsec maximum when C_L = 15 pF; when C_L = 50 pF, t_{PHL} = 5 nsec typical, and no maximum is given.

The computation of propagation delays requires the use of computer-aided methods and is not performed here. However, we estimate the contribution of output transistor Q_5 to t_{PHL}. We use eq. (4.26a) with $\tau = \tau_{ave}$ = 230 psec, approximate $h_{FB} \approx 1$, and ignore all other capacitances except C_L. Thus, the collector current of Q_5 becomes 4.42 mA $\times$ t/230 psec.

Using the considerations of Figure 2.4, the output voltage can be written as V_{out} = 3.9 V - 4.42 mA t^2/(230 psec $\times$ 2 $\times$ C_L). The propagation delay is measured at the +1.5 V level. Thus, we equate V_{out} = 1.5 V, whereby we obtain

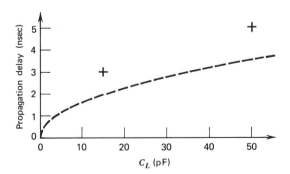

Figure 4.33 Computed propagation delay of the output stage (broken line) and specified values for a HIGH to LOW output transition in a Schottky-diode-clamped TTL gate.

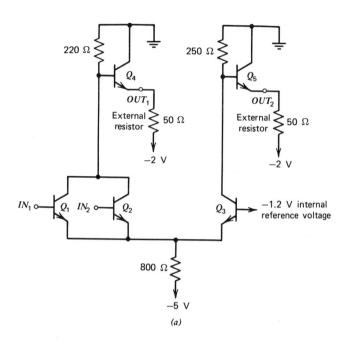

(a)

IN₁ ○——————⊃○ OUT₁
IN₂ ○——————⊃○ OUT₂

(b)

Inputs		Outputs	
IN_1	IN_2	OUT_1	OUT_2
-1.6 V	-1.6 V	-0.8 V	-1.6 V
-1.6 V	-0.8 V	-1.6 V	-0.8 V
-0.8 V	-1.6 V	-1.6 V	-0.8 V
-0.8 V	-0.8 V	-1.6 V	-0.8 V

(c)

Figure 4.34 Emitter-coupled logic (ECL) gate. (a) Circuit diagram; (b) symbol, (c) truth table.

from the above a propagation delay of $t(\text{nsec}) \approx 0.5 \sqrt{C_L(\text{pF})}$. This is shown in Figure 4.33 together with the specifications of the total propagation delay—as expected, the total propagation delay is higher than the propagation delay of the output stage alone.

4.4 EMITTER-COUPLED LOGIC (ECL)

Emitter-coupled logic (ECL) circuits are the fastest commercially available digital circuits. The basic structure of an ECL logic gate is shown in Figure 4.34a. It consists of a current-mode switching circuit (transistors Q_1, Q_2, and Q_3) and of two emitter followers (Q_4 and Q_5). In the following sections, we discuss dc operation of the logic gate and the transient response of the embedded current-mode switching circuit; the operation of emitter followers is discussed in Section 4.2.3.

4.4.1 Basic Operation

Figures 4.35a and 4.35b show approximate operating conditions for the 2-input ECL gate of Figure 4.34 with $h_{FE} \approx 10$. Figure 4.35a shows operation with $IN_1 = IN_2 = -1.6$ V, Figure 4.35b with $IN_1 = -1.6$ V and $IN_2 = -0.8$ V. When

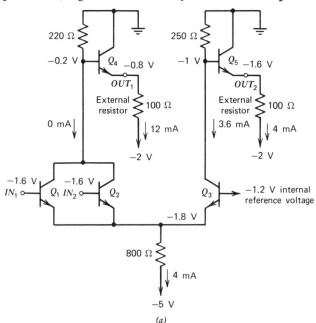

(a)

Figure 4.35 Dc operation of the ECL gate of Figure 4.34 with difference input conditions. (a) $IN_1 = -1.6$ V, $IN_2 = -1.6$ V; (b) $IN_1 = -1.6$ V, $IN_2 = -0.8$ V. (See page 130 for Figure 4.35b.)

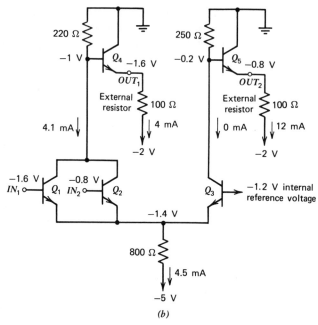

(b)

Figure 4.35 (continued)

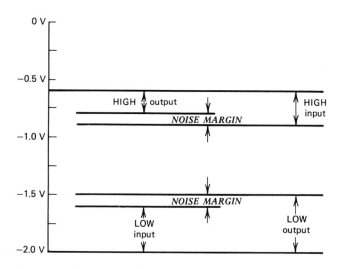

Figure 4.36 Band diagram of noise margins in emitter-coupled logic (ECL) circuits.

the -1.6 V level corresponds to logic 0 and the -0.8 V level to a logic 1, OUT_1 provides a NOR function and OUT_2 an OR function. When the -1.6 V level corresponds to logic 1 and the -0.8 V level to logic 0, OUT_1 provides a NAND function and OUT_2 an AND function.

The ECL circuit shown in Figures 4.34 and 4.35 is only one of the simplest possible. Often the 800 Ω resistor to -5 V is replaced by a current source, and the external 100 Ω resistors to -2 V are replaced by more complex loads.

Noise margins of ECL circuits are shown in Figure 4.36. The margins are smaller than in TTL circuits, however, impedance levels are lower in ECL circuits. With proper interconnections using transmission lines (discussed in Chapter 5), systems of any reasonable size can be constructed.

4.4.2 The Current-Mode Switching Pair

The core of ECL circuits is the current-mode switching pair, shown in Figure 4.37a. Note that the dc current to the common emitters I_{DC} is supplied by a

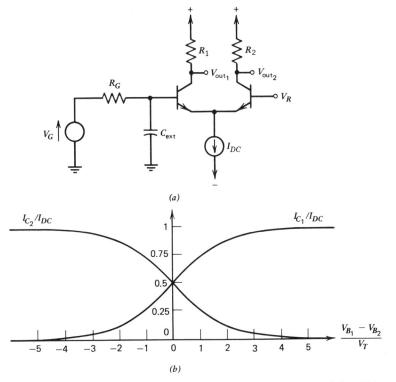

Figure 4.37 Current-mode switching pair. (a) The circuit; (b) basic characteristics; (c) input signal for transient analysis. (See page 132 for Figure 4.37c.)

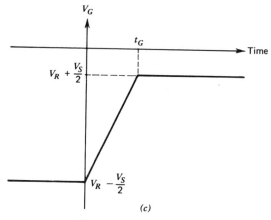

Figure 4.37 (continued)

current source. Also, in what follows we neglect dc base current, that is, we approximate $h_{FE} \approx \infty$.

4.4.2.1 DC Characteristics

Based on eq. (4.1a) we can write for the circuit of Figure 4.37a:

$$I_{E_1} = I_{ES} [e^{(V_{B_1} - V_E)/V_T} - 1],$$ (4.33a)

$$I_{E_2} = I_{ES} [e^{(V_{B_2} - V_E)/V_T} - 1],$$ (4.33b)

and

$$I_{E_1} + I_{E_2} = I_{DC}.$$ (4.33c)

In eqs. (4.33), subscripts 1 refer to the left transistor and subscripts 2 to the right transistor. Also, V_E is the voltage at the emitters and $V_{B_2} = V_R$. Now we make the approximations that $I_{E_1} \gg I_{ES}$ and $I_{E_2} \gg I_{ES}$, which will be seen to be valid. With these approximations, eqs. (4.33) can be combined to result in

$$I_{E_1} = \frac{I_{DC}}{1 + e^{-(V_{B_1} - V_{B_2})/V_T}}$$ (4.34a)

and

$$I_{E_2} = \frac{I_{DC}}{1 + e^{(V_{B_1} - V_{B_2})/V_T}}.$$ (4.34b)

At this point we introduce the *hyperbolic tangent* function, $\tanh(x)$, as follows

$$\tanh(x) \equiv \frac{e^x - e^{-x}}{e^x + e^{-x}}.$$ (4.35)

By use of eq. (4.35), eqs. (4.34) can also be written as

$$I_{E_1} = \frac{I_{DC}}{2}\left[1 + \tanh\left(\frac{V_{B_1} - V_{B_2}}{2V_T}\right)\right],\qquad(4.36a)$$

$$I_{E_2} = \frac{I_{DC}}{2}\left[1 - \tanh\left(\frac{V_{B_1} - V_{B_2}}{2V_T}\right)\right].\qquad(4.36b)$$

Equations (4.36) are equivalent to eqs. (4.35) and are illustrated in Figure 4.37b. When $V_{B_1} = V_{B_2}$, we have $I_{E_1} = I_{E_2} = I_{DC}/2$, as expected. Also, $I_{E_1}/I_{DC} = 0.1$ and $I_{E_2}/I_{DC} = 0.9$ when $V_{B_1} - V_{B_2} = -V_T \ln 9 \approx -55$ mV, and $I_{E_1}/I_{DC} = 0.9$ and $I_{E_2}/I_{DC} = 0.1$ when $V_{B_1} - V_{B_2} = V_T \ln 9 \approx 55$ mV. Thus, most (but not all) of the standing current I_{DC} gets transferred from one transistor to the other one as $V_{B_1} - V_{B_2}$ is swept.

Example 4.16 Various values of I_{E_1}/I_{DC} and I_{E_2}/I_{DC} as functions of $V_{B_1} - V_{B_2}$ are shown in the following table with $V_T = 25$ mV, as computed from eqs. (4.34).

Differential Base Voltage	Normalized Emitter Currents	
$V_{B_1} - V_{B_2}$	I_{E_1}/I_{DC}	I_{E_2}/I_{DC}
-400 mV	1.1×10^{-7}	≈ 1
-200 mV	3.3×10^{-4}	≈ 1
-100 mV	0.018	0.982
-55 mV	0.1	0.9
0	0.5	0.5
55 mV	0.9	0.1
100 mV	0.982	0.018
200 mV	≈ 1	3.3×10^{-4}
400 mV	≈ 1	1.1×10^{-7}

4.4.2.2 Delay and Risetime

In the following we find the delay and the risetime of the collector currents in the circuit of Figure 4.37a for the input signal shown in Figure 4.37c. We approximate the circuit of Figure 4.37a by use of the transistor model of Figure 4.10a with $C_{CB1} = 0$, $C_{CB2} = 0$, and $r_A = \infty$. The result is shown in Figure 4.38. Note that the ohmic base resistances $r_{b_1} = r_{b_2} = r_b$ are now external to the models.

It can be shown that, because of the $h_{FE} \approx \infty$ approximation, $I_{B_1} + I_{B_2} = 0$ in the circuit of Figure 4.38, and that $I_{C_1} = I_{E_1}$ and $I_{C_2} = I_{E_2}$. Also, the charges stored on the bases

$$Q_{BE_1} = \tau I_{C_1}\qquad(4.37a)$$

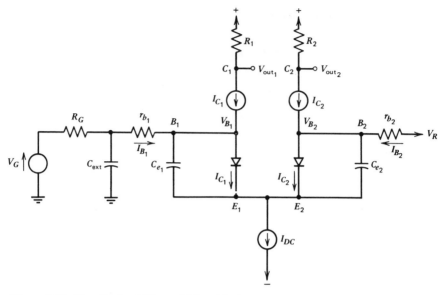

Figure 4.38 The circuit of Figure 4.37*a* with the transistor model of Figure 4.10*a* with $C_{CB_1} = 0$, $C_{CB_2} = 0$, and $r_A = \infty$.

and

$$Q_{BE_2} = \tau I_{C_2}. \tag{4.37b}$$

Hence, Figure 4.37*b* and the table in Example 4.16 also provide $Q_{BE_1}/(\tau I_{DC})$ and $Q_{BE_2}/(\tau I_{DC})$ as functions of $(V_{BE_1} - V_{BE_2})/V_T$. Finally, we note that we use for τ the "average" τ_{ave} given by eq. (4.28).*

In computing the delay and the risetime, we first find the contributions of the input signal, of the external capacitance C_{ext}, and of the nonzero τ of the transistors, and then combine them in approximate expressions for the overall delay and risetime. Finally, we evaluate the effects of capacitive loads on the collectors.

4.4.2.3 Delay and Risetime Originating from Input Risetime t_G

The input signal is shown in Figure 4.37*c*. It has a total voltage swing of V_S that is symmetric around reference voltage V_R—note that this symmetry leads to

---

*This is a very crude approximation. It would be more accurate to remove the contribution of C_{se} from τ_{ave}, and add $C_{se}/2$ to C_{ext}.

delays and risetimes that are identical for positive-going and for negative-going transients. In emitter-coupled logic (ECL) the voltage swing is typically $V_S = 0.8$ V. Transients with $V_S = 0.4$ V are the subject of Problems 16, 18, 19, 20, and 21.

By inspection of Figure 4.37c we can write the delay t_{DG} resulting from the nonzero signal risetime t_G alone as

$$t_{DG} = \frac{t_G}{2}. \tag{4.38}$$

We saw above that the 10% and 90% points of I_{C_1} and I_{C_2} are reached at $V_{B_1} - V_{B_2} = -V_T \ln 9$ and $V_T \ln 9$. Thus, the 10 to 90% risetime t_{RG} resulting from the nonzero signal risetime t_G alone becomes

$$t_{RG} = t_G \frac{2 V_T \ln 9}{V_S}. \tag{4.39}$$

Example 4.17 In an ECL circuit the input voltage swing is $V_S = 0.8$ V, and the risetime of the input signal is $t_G = 1$ nsec; also, $V_T = 25$ mV. Thus, the delay $t_{DG} = 1$ nsec$/2 = 0.5$ nsec and the 10 to 90% risetime $t_{RG} = 1$ nsec $\times 2 \times 25$ mV $(\ln 9)/0.8$ V $= 0.14$ nsec.

4.4.2.4 Delay and Risetime Originating from External Capacitance C_{ext}

In order to find the effects of external capacitance C_{ext} alone, we make $t_G = 0$ and $\tau = 0$; hence, we also have $C_{e_1} = 0$ and $C_{e_2} = 0$. This leaves a simple RC circuit. It can be shown that the resulting delay t_{DC} is

$$t_{DC} = R_G C_{ext} \ln 2, \tag{4.40}$$

and the resulting 10 to 90% risetime t_{RC} is

$$t_{RC} = R_G C_{ext} \ln \frac{1 + (2 V_T \ln 9)/V_S}{1 - (2 V_T \ln 9)/V_S}. \tag{4.41}$$

Note that when $V_S \gg 4 V_T \ln 9 \approx 220$ mV, eq. (4.41) can be approximated as

$$t_{RC} \approx R_G C_{ext} \frac{4 V_T \ln 9}{V_S}. \tag{4.42}$$

Example 4.18 In an ECL circuit $V_S = 0.8$ V, $V_T = 25$ mV, $R_G = 100$ Ω, and $C_{ext} = 1$ pF. The delay t_{DC} from eq. (4.40) is $t_{DC} = 100$ Ω $\times 1$ pF $\times 0.695 \approx 0.07$ nsec. The exact risetime from eq. (4.41) is $t_{RC} = 0.0278$ nsec. The use of the approximate eq. (4.42) leads to $t_{RC} = 0.0275$ nsec.

4.4.2.5 Delay and Risetime Originating from Timeconstant τ

In order to find the effects of nonzero τ alone, we equate $t_G = 0$ and $C_{ext} = 0$. Thus, there remain C_{e_1} and C_{e_2} governing the transient, which is now inherently nonlinear. As a result, unlike the effects of nonzero t_G and C_{ext}, the transient due to nonzero τ can not be determined by linear circuit analysis. In order to find the exact transient the use of computer-aided methods, such as the forward-Euler method described in the preceding chapter, is required. However, approximations for the delay and the risetime may be found from charge conservation considerations. The validity of the approximations will be established by comparison with results of computer-aided analysis (see Reference 5).

We saw that 80% of charges Q_{BE_1} and Q_{BE_2} in eqs. (4.37) are supplied to the bases between $-55 \text{ mV} \leqslant V_{B_1} - V_{B_2} \leqslant 55 \text{ mV}$ and 90% of the charges between $-100 \text{ mV} \leqslant V_{B_1} - V_{B_2} \leqslant 100 \text{ mV}$. Thus, for the purpose of finding the current into the base of Q_1, as a crude approximation we assume $|V_{B_1} - V_{B_2}| \ll V_S/2$. Thus, the current charging the base of Q_1 is approximately $I_{B_1} \approx V_S/[2(R_G + 2r_b)]$.

In computing the delay of the 50% point $t_{D\tau}$ we note that a charge of $\tau I_{DC}/2$ must be supplied to the base of Q_1, whereby $t_{D\tau} = \tau I_{DC}/(2I_{B_1})$, leading to

$$t_{D\tau} \approx \tau \frac{I_{DC}(R_G + 2r_b)}{V_S}. \tag{4.43}$$

It can be shown that a more accurate approximation of $t_{D\tau}$ is given by

$$t_{D\tau} = \tau \frac{I_{DC}(R_G + 2r_b)}{V_S}\left(1 - \frac{2V_T}{V_S}\right). \tag{4.44}$$

Computer-aided analysis shows that eq. (4.44) is accurate within 10% when $V_S \geqslant 6 \ V_T \approx 150 \text{ mV}$ and within 5% when $V_S \geqslant 16 \ V_T \approx 400 \text{ mV}$. Note that the cruder eq. (4.43) leads to an added fractional error of at most $2V_T/V_S$, which is only 6.25% for $V_S = 0.8 \text{ V}$.

In computing the risetime $t_{R\tau}$ we note that a charge of $0.8 \ I_{DC}\tau$ has to be supplied to the base of Q_1. This results in

$$t_{R\tau} \approx 1.6 \frac{I_{DC}(R_G + 2r_b)}{V_S}. \tag{4.45}$$

Computer-aided analysis shows that eq. (4.45) is accurate within 5% when $V_S \geqslant 16 \ V_T \approx 400 \text{ mV}$. (A more accurate approximation for $t_{R\tau}$ is given in Problem 22.)

Example 4.19 In a current-mode switching pair $\tau = 0.2$ nsec, $r_b = 50$ Ω, $R_G = 100$ Ω, $V_S = 0.8$ V, and $V_T = 25$ mV. By use of the approximate eq. (4.43) we get a delay of $t_{D\tau} \approx 0.5$ nsec, by use of the more accurate eq. (4.44) a $t_{D\tau} = 0.47$ nsec. Also, eq. (4.45) yields a 10 to 90% risetime of $t_{R\tau} \approx 0.8$ nsec.

4.4.2.6 Simultaneous Effects of t_G, C_{ext}, and τ

When the effects of nonzero t_G, C_{ext}, and τ must be taken into account simultaneously, we have to resort to computer-aided analysis. However, crude approximations for the resulting delay t_D and the resulting risetime t_R are also available.

The resulting delay t_D can be approximated by simply adding t_{DG}, t_{DC}, and $t_{D\tau}$—similarly to what was done for the linear case in Section 2.11. Thus,

$$t_D \approx t_{DG} + t_{DC} + t_{D\tau}, \tag{4.46}$$

where t_{DG} is given by eq. (4.38), t_{DC} by eq. (4.40), and $t_{D\tau}$ by eq. (4.43) or eq. (4.44). Equation (4.46) provides only a very crude approximation. Computer-aided analysis shows that it may be inaccurate by as much as 35% when the three terms in it are comparable with each other.

The resulting risetime t_R can be approximated as

$$t_R = \sqrt{t_{RG}^2 + (t_{RC} + t_{R\tau})^2}, \tag{4.47}$$

where t_{RG} is given by eq. (4.39), t_{RC} by eq. (4.41) or eq. (4.42), and $t_{R\tau}$ by eq. (4.45). The accuracy of eq. (4.47) is limited primarily by the accuracies of t_{RC} and $t_{R\tau}$.

Example 4.20 The combination of the delays from Examples 4.17, 4.18, and 4.19 results in a delay of $t_D \approx 0.5$ nsec + 0.07 nsec + 0.47 nsec = 1.04 nsec. The combination of the risetimes from Examples 4.17, 4.18, and 4.19 results in a risetime of

$$t_R \approx \sqrt{(0.14 \text{ nsec})^2 + (0.0278 \text{ nsec} + 0.8 \text{ nsec})^2} = 0.84 \text{ nsec}.$$

4.4.2.7 Effects of a Capacitive Load at the Output

The delay t_D and the risetime t_R computed thus far are those of I_{C_1} and I_{C_2}. However, often we are interested in the delay and the risetime of the output voltage when the output is loaded by a capacitance C_L. The resulting delay $t_{D\,\text{out}}$ and the resulting 10 to 90% risetime $t_{R\,\text{out}}$ can be obtained by the methods of Section 2.11 as

$$t_{D\,\text{out}} = t_D + RC_L \tag{4.48}$$

and

$$t_{R_{\text{out}}} = \sqrt{t_R^2 + 2\pi(RC_L)^2}, \tag{4.49}$$

where t_D is given by eq. (4.46), t_R by eq. (4.47), and R equals R_1 or R_2, whichever is applicable.

Example 4.21   In the circuit of Example 4.20 the load resistors $R_1 = R_2 = 100 \ \Omega$ and the capacitances loading R_1 and R_2 can each be approximated as $C_L = 2$ pF. Thus, for either output, the overall delay

$$t_{D_{\text{out}}} = 1.04 \text{ nsec} + 0.2 \text{ nsec} = 1.24 \text{ nsec}$$

and the overall risetime

$$t_{r_{\text{out}}} = \sqrt{(0.84 \text{ nsec})^2 + 2\pi(0.2 \text{ nsec})^2} = 0.98 \text{ nsec}.$$

PROBLEMS

1.   Derive eq. (4.2) from eq. (4.1).

2.   The current gain of an NPN transistor is $h_{FE} = 100$. What are the values of α, $I_{C_{DC}}$, and $I_{B_{DC}}$ if $I_{E_{DC}} = 20$ mA?

3.   Summarize in a table the four operating regions of an NPN transistor. Use the voltage and current conventions of Figure 4.3. Include the states (forward biased or reverse biased) of both junctions in each of the four regions.

4.   A high speed transistor is characterized by $\tau_{de} = 100$ psec and $I_{ES} = 1$ pA. The transistor is operated at a dc emitter current of $I_{E_{DC}} = 20$ mA. Find the base-emitter diffusion capacitance C_{de} and the stored base-emitter charge Q_{BE_d} at room temperature. Assume that C_{te} and C_{se} are both negligible.

5.   Find C_{se}/C_{de} and Q_{BE_s}/Q_{BE_d} for the transistor described in Examples 4.4 and 4.5 operated at $I_{E_{DC}} = 1$ mA.

6.   How does the position of the curve in Figure 4.9 change if τ_{de} is changed from 100 psec to 200 psec and C_{se} and V_T remain unchanged.

[†]7.   Use the results of Figure 2.45 and sketch the voltage across R_L in Figure 4.13 with the conditions of Example 4.7(a), if $R_L = 200 \ \Omega$ and if a capacitance of 0.5 pF is connected in parallel with R_L.

[†]8.   Use eq. (4.8) and demonstrate that the small-signal output resistance r_{ce} due to the Early effect is $r_{ce} = V_A/(h_{FE}I_B)$, where V_A is the Early voltage.

[†]9.   An expression for the output resistance due to the Early effect was given in Problem 8. Use this expression and find the voltage across R_L in Figure 4.13 with the conditions of Example 4.7(a) if $R_L = 10$ kΩ and if the Early voltage is (a) $V_A = \infty$, (b) $V_A = 10$ V.

10. Find the component values in Figure 4.14 and in Figure 4.15 for the transistor described in Example 4.7 if $r_b = 50\ \Omega$.

11. In the circuit of Figure 4.17, the small-signal part of I_B is $i_{b_0} \sin (2\pi ft)$. Demonstrate that

$$|i_c| = h_{FE} |i_{b_0}|/\sqrt{1 + [(h_{FE} + 1)f/f_T]^2}$$

and

$$\underline{/i_c} = -\arctan [(h_{FE} + 1)f/f_T]$$

12. A transistor is characterized by $\tau_{de} = 200$ psec, $C_{se} = 1$ pF, $I_{ES} = 0.1$ pA, and $V_T = 25$ mV. Find τ_{ave} if the transistor is switched between (a) $I_{E_1} = 0$ and $I_{E_2} = 20$ mA, (b) $I_{E_1} = 20\ \mu$A and $I_{E_2} = 20$ mA.

13. In the circuit of Example 4.15, a resistance of $R_G = 400\ \Omega$ is added in series with inductance L.

   (a) Find the minimum value of load capacitance C_L that is required to ensure that the circuit is stable with any positive L.

   (b) Find the maximum permitted value of L if $C_L = 3$ pF and if a stable circuit is desired.

14. (a) Derive eqs. (4.34),

    (b) Derive eqs. (4.36) from eqs. (4.34) and (4.35).

15. At what value of $V_{B_1} - V_{B_2}$ is $I_{C_1}/I_{DC} = 0.01$ in the table of Example 4.16?

16. Repeat Example 4.17 with $V_S = 0.4$ V.

17. Derive eq. (4.41) and eq. (4.42).

18. Repeat Example 4.18 with $V_S = 0.4$ V.

19. Repeat Example 4.19 with $V_S = 0.4$ V.

20. Repeat Example 4.20 with $V_S = 0.4$ V.

21. Repeat Example 4.21 with $V_S = 0.4$ V.

†22. Show analytically that in a current-mode switching pair

$$t_{D\tau} = \frac{R_G + 2r_b}{V_S}\ I_{DC}\tau \left[1 - 2\frac{V_T}{V_S} \pm \cdots \right]$$

and

$$t_{R\tau} = 1.6\frac{R_G + 2r_b}{V_S}\ I_{DC}\tau \left[1 + \left(\frac{1.85\ V_T}{V_S}\right)^2 + \cdots \right].$$

Use Taylor series expansions as indicated in the Appendix of Reference 5.

REFERENCES

1. J. Millman and H. Taub, *Pulse, Digital, and Switching Waveforms*, McGraw-Hill, New York, 1965.

2. S. D. Senturia and B. D. Wedlock, *Electronic Circuits and Applications*, John Wiley and Sons, New York, 1975.

3. A. Barna, *High-Speed Pulse Circuits*, Wiley-Interscience, New York, 1970.

4. A. Barna and H. Horn, "Instabilities in Emitter Followers and Differential Pairs," Archiv für Elektronik und Übertragungstechnik–Electronics and Communication, **32**, 446–449 (1978).

5. A. Barna, "Delay and Rise Time in Current-Mode Switching Circuits," Archiv für Elektronik und Übertragungstechnik–Electronics and Communication, **30**, 112–116 (1976).

CHAPTER **5**

TRANSMISSION LINES

In the preceding chapters high speed characteristics of an interconnection were simply approximated by a capacitance to ground. This approximation, however, becomes inaccurate when the length of an interconnection is not negligible compared to the product of the signal risetime and the speed of light. In such cases the interconnection must be treated as a *transmission line*.

The treatment presented here focuses on uniform transmission lines. Also, initially all losses are assumed negligible; they are, however, examined in the last section. Transients for linear resistive and capacitive terminations are described by use of classical transmission line theory; these provide satisfactory tools for the treatment of interconnections between emitter-coupled logic (ECL) circuits. Transistor-transistor logic (TTL) circuits, however, present nonlinear resistive terminations to interconnecting lines; these are treated by the graphical method.

The majority of the examples in this chapter utilize coaxial cables as transmission lines, but most figures depict them as wire-pairs—for ease of drawing. All discussion, however, is applicable to all line configurations unless otherwise stated.

5.1 BASIC PROPERTIES

Consider a section of a transmission line, shown in Figure 5.1. Points **A** and **B** along the line are observed by measuring the voltages between the two lines and the currents in the upper line. Let us assume that these have been constant, thus

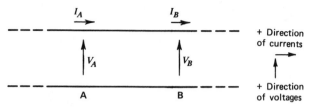

Figure 5.1 A section of a transmission line.

$V_{A\text{initial}} = V_{B\text{initial}}$ and $I_{A\text{initial}} = I_{B\text{initial}}$. When a *signal*, represented by *changes* in the voltage and current in the line, is *traveling* from left to right, V_A and I_A will change first, followed by changes in V_B and I_B. Similarly, when a signal is traveling from right to left, V_B and I_B will change first, and V_A and I_A will change somewhat later. It is also possible to have a composite of two signals traveling in opposite directions: this case can be treated by a superposition of the two signals.*

Let us designate the *changes* (signals) in the voltages and currents by the lower case letters v_A, i_A, v_B, and i_B as follows:

$$V_A = V_{A\text{initial}} + v_A, \tag{5.1a}$$

$$I_A = I_{A\text{initial}} + i_A, \tag{5.1b}$$

$$V_B = V_{B\text{initial}} + v_B, \tag{5.1c}$$

$$I_B = I_{B\text{initial}} + i_B; \tag{5.1d}$$

and similarly for any point along the line:

$$V = V_{\text{initial}} + v, \tag{5.1e}$$

$$I = I_{\text{initial}} + i. \tag{5.1f}$$

Example 5.1 The transmission line of Figure 5.1 is driven on its left end by the output of an emitter-coupled logic circuit. Initially the output voltage of the emitter-coupled logic circuit and voltages $V_{A\text{initial}}$ and $V_{B\text{initial}}$ are at -1.6 V. When the output of the emitter-coupled logic circuit changes from -1.6 V to -0.8 V, it initiates a change of $+0.8$ V along the line. Thus V_A will change by $v_A = +0.8$ V from -1.6 V to -0.8 V, and V_B will change by $v_B = +0.8$ V from -1.6 V to -0.8 V.

The possibility of a signal traveling along a transmission line ("traveling wave"*) is supported by differential equations of transmission line theory. See, for example, W. C. Johnson, *Transmission Lines and Networks*, McGraw-Hill, New York, 1950, page 8.

5.1.1 Characteristic Impedance

Simple relationships between signals v_A, i_A, v_B, and i_B are provided by transmission line theory for a signal traveling in one direction: to the right *or* to the left. In such a case v_A/i_A, v_B/i_B, and the velocity of signal propagation are independent of the signal itself and are characteristic of the transmission line. With the positive directions of the voltages and currents shown in Figure 5.1, for a signal traveling *from the left to the right*:

$$\frac{v_A}{i_A} = Z_0, \qquad (5.2a)$$

$$\frac{v_B}{i_B} = Z_0, \qquad (5.2b)$$

and similarly for any point along the line

$$\frac{v}{i} = Z_0. \qquad (5.2c)$$

With the positive directions of voltages and currents shown in Figure 5.1, for a signal traveling *from the right to the left*:

$$\frac{v_A}{i_A} = -Z_0, \qquad (5.3a)$$

$$\frac{v_B}{i_B} = -Z_0, \qquad (5.3b)$$

and similarly for any point along the line

$$\frac{v}{i} = -Z_0. \qquad (5.3c)*$$

The quantity Z_0 has a dimension of Ω and is designated the *characteristic impedance*, or *characteristic resistance*, of the transmission line. In high speed digital systems, Z_0 is typically between 30 Ω and 200 Ω.

---

*It is tempting to define opposite current directions for the signals traveling to the right and to the left: such a choice could eliminate the negative signs in eqs. (5.3a) through (5.3c). Correct electrical circuit analysis, however, demands that once positive directions are chosen in a circuit, they must be adhered to.

5.1.2 Velocity of Signal Propagation

The velocity of signal propagation is characteristic of the transmission line and is always below that of light in vacuum, c = 30 cm/nanosecond $\approx$ 1 foot/nanosecond. When the dielectric of the transmission line is vacuum or air, the resulting velocity of signal propagation v is near c, although it is still below c. In general, the velocity of signal propagation

$$v = \frac{c}{\sqrt{\epsilon_r}} \text{ (m/sec)} \leqslant c, \tag{5.4}$$

where $\epsilon_r \geqslant 1$ is the *relative dielectric constant*, or *permittivity*, of the dielectric, and c is in m/sec.

Example 5.2 The dielectric of a transmission line is solid polyethylene with a relative dielectric constant of ϵ_r = 2.25. Thus the velocity of signal propagation in the transmission line is

$$v = \frac{c}{\sqrt{\epsilon_r}} = \frac{30 \text{ cm/nsec}}{\sqrt{2.25}} = 20 \text{ cm/nsec} \approx 8 \text{ in./nsec.}$$

When the dielectric of the transmission line is composed of two materials with different ϵ_r's, the calculation of the velocity of signal propagation v becomes more involved. However, it can always be said that the resulting v is between the values obtained for each of the two ϵ_r's.

Example 5.3 The dielectric of the RG 62 A/U coaxial cable is partly polyethylene with ϵ_r = 2.25, partly air. The velocity of signal propagation is v = 0.84 $c \approx$ 25 cm/nsec. This value of v is less than the velocity of signal propagation in air, which is 30 cm/nsec; it is, however, above the velocity of signal propagation in solid polyethylene, which is 20 cm/nsec (see Example 5.2).

5.1.3 Propagation Delay, Capacitance, and Inductance

Given the velocity of signal propagation v of a transmission line in m/sec, its *propagation delay*, T_0, can be written as

$$T_0 = \frac{l}{v} \text{ (sec)}, \tag{5.5}$$

where l is the length of the transmission line in m. By utilizing eq. (5.4), eq. (5.5) may be also written as

$$T_0 = \frac{\sqrt{\epsilon_r}}{c} \, l \, (\text{sec}). \tag{5.6}$$

Assume that we take a transmission line characterized by a propagation delay T_0 and a characteristic impedance Z_0; we leave one end open-circuited, and we measure its *capacitance* at the other end. If the capacitance measuring instrument uses a frequency that is much lower than $1/T_0$, then the measured capacitance C_0 is

$$C_0 = \frac{T_0}{Z_0} \, (\text{Farad}), \tag{5.7}$$

or, by utilizing eq. (5.6),

$$C_0 = \frac{\sqrt{\epsilon_r}}{cZ_0} \, l \, (\text{Farad}). \tag{5.8}$$

Equation (5.8) provides the capacitance of a transmission line with length l, whence the *capacitance per unit length* is $\sqrt{\epsilon_r}/(cZ_0)$ in Farad/m.

The *inductance* L_0 of a transmission line can be measured by short-circuiting one end and measuring the inductance at the other end by use of a frequency $\ll 1/T_0$. This results in

$$L_0 = Z_0 T_0 \, (\text{Henry}), \tag{5.9}$$

or, by utilizing eq. (5.6),

$$L_0 = Z_0 \frac{\sqrt{\epsilon_r}}{c} \, l \, (\text{Henry}). \tag{5.10}$$

The *inductance per unit length* is $Z_0 \sqrt{\epsilon_r}/c$ in Henry/m.

Example 5.4 A transmission line has a characteristic impedance of $Z_0 = 50 \, \Omega$ and a solid dielectric with an $\epsilon_r = 2.25$. Thus the capacitance per unit length is

$$\frac{\sqrt{\epsilon_r}}{cZ_0} = \frac{\sqrt{2.25}}{3 \times 10^8 \text{ m/sec} \times 50 \, \Omega} = 100 \text{ pF/m} \approx 30 \text{ pF/foot}.$$

The inductance per unit length is

$$\frac{Z_0 \sqrt{\epsilon_r}}{c} = \frac{50 \, \Omega \, \sqrt{2.25}}{3 \times 10^8 \text{ m/sec}} = 0.25 \, \mu\text{H/m} \approx 0.075 \, \mu\text{H/foot}.$$

5.1.4 Transmission Line Configurations

One of the most common transmission line configurations in use in digital electronics is a wire above the ground plane shown in Figure 5.2. Interconnec-

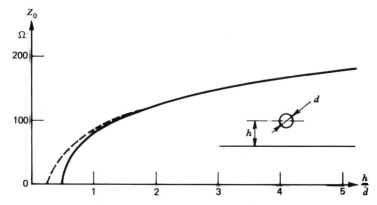

Figure 5.2 A wire above the ground plane and its characteristic impedance for $\epsilon_r = 1$. The solid line represents the exact $Z_0 = 60$ arccosh $2h/d$; the broken line represents the approximation $Z_0 = 60 \ln 4h/d$ valid for $h/d \gg 1$.

tions are frequently made also by coaxial cable shown in Figure 5.3: d_1 is the outer diameter of the inner conductor, d_2 is the inner diameter of the outer conductor. When $\epsilon_r \neq 1$, Z_0 must be divided by $\sqrt{\epsilon_r}$.

Example 5.5 An RG 58 B/U coaxial cable has a solid polyethylene dielectric ($\epsilon_r = 2.25$) with an outer diameter of 0.116 in. ≈ 3 mm. The outer conductor covers the dielectric without a significant gap, hence $d_2 = 3$ mm. The center conductor is a #20 wire, $d_1 = 0.032$ in. ≈ 0.81 mm. Thus $d_2/d_1 = 3$ mm/0.81 mm ≈ 3.7. If the dielectric were air ($\epsilon_r = 1$), then for $d_2/d_1 = 3.7$ we would get from Figure 5.3 a $Z_0 = 80.5$ Ω. For the polyethylene dielectric with $\epsilon_r = 2.25$, $Z_0 = 80.5$ $\Omega/\sqrt{2.25} = 53.5$ Ω.

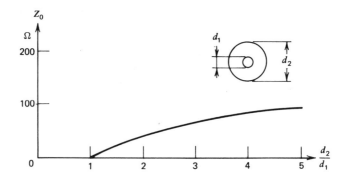

Figure 5.3 The coaxial line and its characteristic impedance for $\epsilon_r = 1$: $Z_0 = 60 \ln d_2/d_1$.

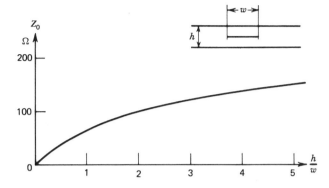

Figure 5.4 The stripline and its characteristic impedance for $\epsilon_r = 1$.

Two other transmission line configurations that find extensive use in digital systems are the stripline and the microstrip. The stripline, shown in Figure 5.4, is used for interconnections within multilayer boards and substrates; the microstrip, shown in Figure 5.5, is used on the top layers.

Note that all characteristic impedances are given for $\epsilon_r = 1$. When $\epsilon_r \neq 1$, Z_0 must be divided by $\sqrt{\epsilon_r}$ to get the correct value. Finding the ϵ_r to be used may be difficult when the dielectric is composed of two different materials: such is the case for the microstrip of Figure 5.5 where part of the dielectric has a dielectric constant of ϵ_r and part of it is air, since the stray electric field extends into the air. As a rough approximation, however, an *effective relative dielectric constant* ϵ_{eff} can be found for the microstrip as

$$\epsilon_{\text{eff}} \approx (1 + \epsilon_r)/2, \tag{5.11}$$

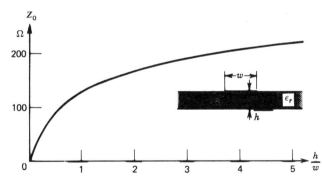

Figure 5.5 The microstrip and its characteristic impedance for $\epsilon_r = 1$.

and the correct characteristic impedance is obtained by dividing Z_0 of Figure 5.5 by $\sqrt{\epsilon_{\text{eff}}}$. This approximation is accurate within about 10% when the resulting characteristic impedance is between 50 Ω and 150 Ω – a practical range in high speed digital systems.* Velocity of signal propagation, capacitance, and inductance can also be found by using ϵ_{eff} instead of ϵ_r in eqs. (5.4), (5.8), and (5.10).

5.2 LINEAR RESISTIVE TERMINATION

Thus far we have considered signal propagation in a transmission line without regard to what takes place at its ends. Either end may be *terminated* by another (possibly identical) transmission line, a resistor or other impedance, or by any other two-pole linear or nonlinear network.

5.2.1 Termination by the Characteristic Impedance

One of the simplest cases is shown in Figure 5.6, where the right (load) end of the transmission line is terminated by a resistance R_L with a value that is equal to the characteristic impedance Z_0 of the transmission line. With the signs of voltages and currents shown, load resistor R_L imposes at point **L** a $V_L/I_L = Z_0$. This holds for initial conditions as well as for changes (signals) in eqs. (5.1), that is, $V_{L_{\text{initial}}}/I_{L_{\text{initial}}} = R_L = Z_0$, and $v_L/i_L = R_L = Z_0$. Thus, by comparing with eq. (5.2c), we can see that the load resistor looks just like a further length of identical transmission line, hence no *reflection* takes place when a signal traveling from the left to the right is incident at point **L**.

The transients are illustrated in the circuit of Figure 5.7, where a transmission line with characteristic impedance Z_0 and propagation delay T_0 is driven at point **G** by a voltage source (generator) and is terminated at point **L** by a load

*H. R. Kaupp, "Characteristics of Microstrip Transmission Lines," *IEEE Transactions on Electronic Computers, EC-16*, No. 2, 185–193 (April 1967).

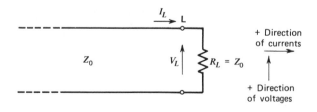

Figure 5.6 Transmission line terminated by its characteristic impedance.

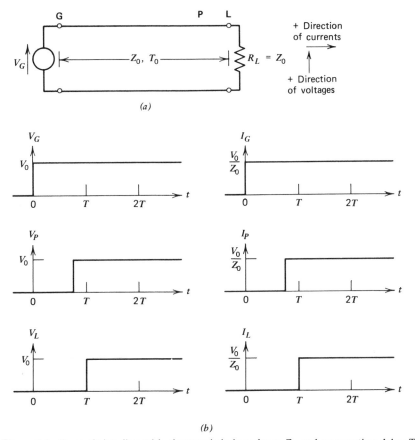

Figure 5.7 Transmission line with characteristic impedance Z_0 and propagation delay T_0 driven by a voltage source and terminated by a load resistor $R_L = Z_0$. (*a*) The circuit; (*b*) transients for a step voltage input.

resistance R_L that has a value equal to Z_0. At time $t = 0$, the voltage source changes from $V_G = 0$ to $V_G = V_0$ (a *step voltage*), and a signal is initiated with a voltage of V_0 traveling to the right. Since for a signal traveling to the right the ratio of the voltage to the current is Z_0, the current at point **G** changes from zero to V_0/Z_0 at time $t = 0$. The signal traveling to the right arrives at the inter-mediate point **P** approximately at time $t = 0.75\ T_0$ (signals V_P and I_P), and at the right end point **L** at time $t = T_0$ (signals V_L and I_L). Since no reflection takes place, no further changes occur in the voltages or in the currents.

Example 5.4 An RG 58 A/U coaxial cable has a characteristic impedance of $Z_0 = 50\ \Omega$, a length of 10 feet (≈ 3 m), and its dielectric is solid poly-ethylene with $\epsilon_r = 2.25$. Thus, from eq. (5.6),

$$T_0 = \frac{\sqrt{\epsilon_r}}{c}\, l = \frac{\sqrt{2.25}}{3 \times 10^8 \text{ m/sec}}\, 3 \text{ m} = 15 \text{ nanoseconds.}$$

This coaxial cable is used as a transmission line in the circuit of Figure 5.7 with $V_0 = 1$ V. Thus the steps in the currents have magnitudes of $V_0/Z_0 = 1$ V/50 Ω = 20 mA, and the signal reaches point **L** at a time $t = T_0 = 15$ nanoseconds.

5.2.2 Termination by an Open Circuit

Consider next termination by an open circuit, as shown in Figure 5.8. Let us assume that a signal traveling from the left to the right is incident at point **L** and that it is characterized by an *incident voltage* v_i and an *incident current* i_i. Thus, based on eq. (5.2c),

$$\frac{v_i}{i_i} = Z_0. \tag{5.12a}$$

When this signal arrives at point **L**, current i_i has no place to go, hence an additional current component, a *reflected current* i_r, appears as part of I_L such that

$$i_i + i_r = I_L = 0. \tag{5.12b}$$

Also, an additional voltage component, *reflected voltage* v_r, appears as part of V_L such that

$$v_i + v_r = V_L. \tag{5.12c}$$

Reflected current i_r and reflected voltage v_r will now start traveling to the left. However, for a signal traveling from the right to the left, based on eq. (5.3c),

$$\frac{v_r}{i_r} = -Z_0. \tag{5.12d}$$

By combining eqs. (5.12a), (5.12b), and (5.12d), we get for the *reflection coefficient* Γ, defined as the ratio of v_r to v_i,

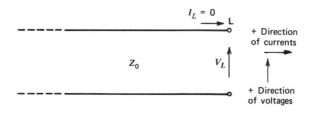

Figure 5.8 Transmission line terminated by an open circuit.

$$\Gamma = \frac{v_r}{v_i} = 1; \qquad (5.13a)$$

also,

$$\frac{i_r}{i_i} = -1 = -\Gamma. \qquad (5.13b)$$

Hence, for a transmission line terminated by an open circuit, the reflected voltage is identical to the incident voltage and the reflected current is opposite to the incident current.

The transients are illustrated in the circuit of Figure 5.9a where a transmission line with characteristic impedance Z_0 and propagation delay T_0 is driven by a voltage source in series with a resistor R_G that has a value equal to Z_0. Signals for a step voltage input are shown in Figure 5.9b. At time $t = 0$ the voltage source changes from $V_S = 0$ to $V_S = V_0$, the voltage at point **G** changes from $V_G = 0$ to $V_G = V_0/2$, and a signal traveling to the right is initiated. Since for a signal traveling to the right the ratio of the voltage to the current is Z_0, the current at point **G** changes from zero to $V_0/(2Z_0)$ at time $t = 0$.

The signal traveling to the right arrives at intermediate point **P** at time $t = 0.75T_0$ (signals V_P and I_P), and it arrives at the open-circuited right end point **L** at time $t = T_0$. This incident signal is characterized by a voltage of $v_i = V_0/2$ and a current of $i_i = V_0/(2Z_0)$. According to eq. (5.13a), the reflected voltage $v_r = v_i = V_0/2$, resulting in a total voltage of $V_L = v_i + v_r = V_0/2 + V_0/2 = V_0$ at time $t = T_0$, as shown in the bottom left graph of Figure 5.9b. Also, the reflected current, from eq. (5.13b), is $i_r = -i_i = -V_0/(2Z_0)$, hence the total current $I_L = i_i + i_r = V_0/(2Z_0) - V_0/(2Z_0) = 0$ as expected.

The reflected signal characterized by $v_r = V_0/2$ and $i_r = -V_0/(2Z_0)$ travels to the left and reaches intermediate point **P** at time $t = 1.25T_0$. Voltage $v_r = V_0/2$ adds to the already present voltage of $V_0/2$, resulting in a total voltage of V_0. Similarly, current $i_r = -V_0/(2Z_0)$ adds to the already present current $V_0/(2Z_0)$ resulting in zero total current. Finally, the reflected signal reaches point **G** at time $t = 2T_0$: the line is terminated here by an $R_G = Z_0$, thus no further reflections will take place.

The foregoing discussion deals with transients for a step voltage input illustrated in Figure 5.9b. When the input is a *pulse* that can be treated as a sum (superposition) of two step voltages, each signal in the circuit can be found as a sum of the individual transients resulting from the two step voltages.

Example 5.5 In the circuit of Figure 5.9a, input signal V_S is a pulse with a height of V_0 and a duration of $3T_0$, as shown in the upper part of Figure 5.9c. This pulse can be treated as a sum of two step voltages: one of these has a height of V_0 and starts at $t = 0$, the other one has a height of $-V_0$ and starts at $t = 3T_0$.

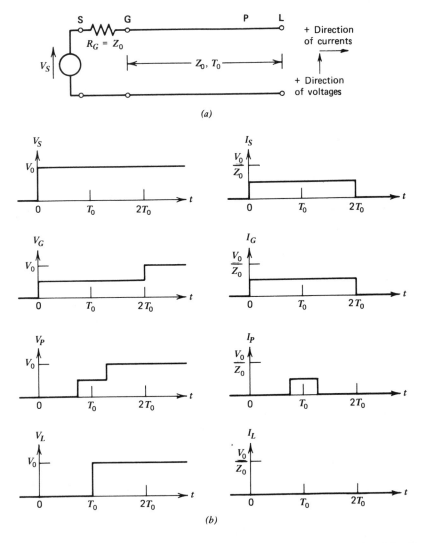

Figure 5.9 Transmission line with characteristic impedance Z_0 and propagation delay T_0 driven by a voltage source in series with a resistor $R_G = Z_0$ and terminated by an open circuit. (a) The circuit; (b) transients for a step voltage input; (c) transients for an input pulse with a width of 3 T_0. (See page 153 for Figure 5.9c.)

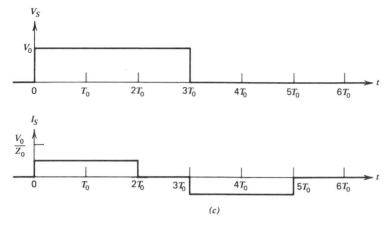

Figure 5.9 (continued)

The resulting I_S is shown in the lower part of Figure 5.9c. It was obtained as the sum of two signals: one of these is I_S copied from Figure 5.9b, the other one was inverted and also delayed by $3T_0$.

Note that although the input voltage pulse is positive, upon its termination a negative source current I_S results that must be carried by the voltage source.

5.2.3 Termination by a Short Circuit

Consider next a transmission line terminated by a short circuit, shown in Figure 5.10. A signal with a voltage v_i and current i_i traveling from the left to the right is characterized by

$$\frac{v_i}{i_i} = Z_0. \tag{5.14a}$$

When this signal is incident at point **L**, a reflected voltage v_r appears, making the

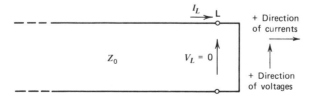

Figure 5.10 Transmission line terminated by a short circuit.

total voltage V_L zero:

$$v_i + v_r = V_L = 0 \qquad (5.14b)$$

and a reflected current i_r appears, determined by

$$i_i + i_r = I_L. \qquad (5.14c)$$

The reflected signal, characterized by v_r and i_r, now starts traveling from point L to the left. However, for a signal traveling from the right to the left, based on eq. (5.3c),

$$\frac{v_r}{i_r} = -Z_0. \qquad (5.14d)$$

By combining eqs. (5.14a), (5.14b), and (5.14d), we get for the reflection coefficient

$$\Gamma = \frac{v_r}{v_i} = -1; \qquad (5.15a)$$

also,

$$\frac{i_r}{i_i} = 1 = -\Gamma. \qquad (5.15b)$$

The transients for short circuit termination are illustrated in the circuit of Figure 5.11a. Figure 5.11b shows the transients for a step voltage input, Figure 5.11c for an input pulse with a width of $3T_0$.

5.2.4 Termination by an Arbitrary Resistance

When the transmission line is terminated by a resistance R_L that may have any value (Figure 5.12), we can say that the ratio of V_L to I_L has to be R_L:

$$\frac{V_L}{I_L} = R_L. \qquad (5.16a)$$

Also, from previous considerations,

$$V_L = v_i + v_r, \qquad (5.16b)$$

$$I_L = i_i + i_r, \qquad (5.16c)$$

$$\frac{v_i}{i_i} = Z_0, \qquad (5.16d)$$

$$\frac{v_r}{i_r} = -Z_0. \qquad (5.16e)$$

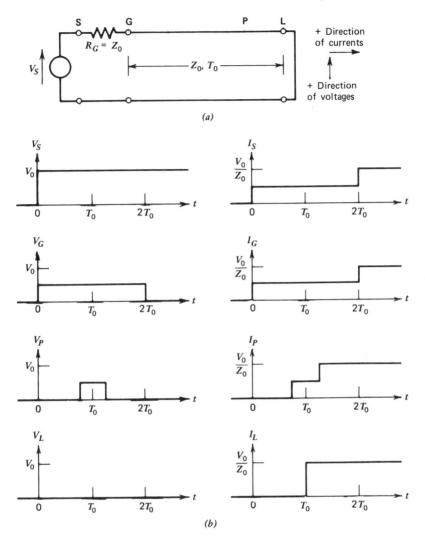

Figure 5.11 Transmission line with characteristic impedance Z_0 and propagation delay T_0 driven by a voltage source in series with a resistor $R_G = Z_0$ and terminated by a short circuit. (a) The circuit; (b) transients for a step voltage input; (c) transients for an input pulse with a width of $3T_0$. (See page 156 for Figure 5.11c.)

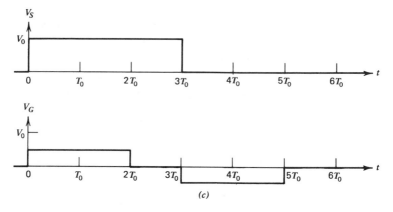

Figure 5.11 (continued)

From eqs. (5.16) we can obtain the reflection coefficient Γ as

$$\Gamma = \frac{v_r}{v_i} = \frac{R_L - Z_0}{R_L + Z_0}; \qquad (5.17a)$$

also

$$\frac{i_r}{i_i} = -\frac{R_L - Z_0}{R_L + Z_0} = -\Gamma. \qquad (5.17b)$$

We can also see that when $R_L \to \infty$, eqs. (5.17) reduce to eqs. (5.13), when $R_L = 0$ they reduce to eqs. (5.15), and when $R_L = Z_0$ then the reflection coefficient $\Gamma = 0$.

When neither end of the transmission line is terminated by its characteristic impedance Z_0 (see Figure 5.13a), multiple reflections will take place. In such a case separate reflection coefficients Γ_G and Γ_L can be defined, respectively, for the left (generator) and right (load) ends of the transmission line as

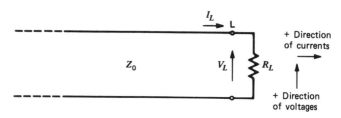

Figure 5.12 Transmission line terminated by an arbitrary resistance R_L.

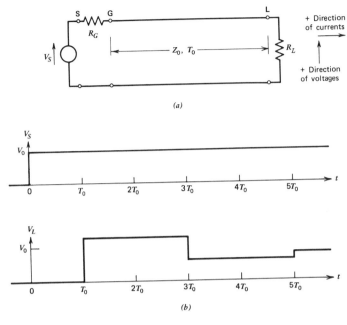

(a)

(b)

Figure 5.13 Transmission line with characteristic impedance Z_0 and propagation delay T_0 driven by a voltage source in series with a resistor R_G and terminated by a resistor R_L. (a) The circuit; (b) voltage V_L for $R_G = 0.2Z_0$ and $R_L = 4Z_0$.

$$\Gamma_G = \frac{R_G - Z_0}{R_G + Z_0} \qquad (5.18a)$$

and

$$\Gamma_L = \frac{R_L - Z_0}{R_L + Z_0}. \qquad (5.18b)$$

By use of eqs. (5.18), V_L can be found for a step voltage input of V_0 as*

$$V_L = V_0 \frac{R_L}{R_G + R_L}[1 - (\Gamma_G \Gamma_L)^N] \qquad \text{for } t > T_0 \qquad (5.19a)$$

and as

$$V_L = 0 \qquad \text{for } t < T_0, \qquad (5.19b)$$

*See, for example, A. Barna, *High-Speed Pulse Circuits*, Wiley-Interscience, New York, 1970, p. 37.

where $N = 1, 2, 3, \ldots$ defines the range of applicable t as

$$(2N - 1)T_0 < t < (2N + 1)T_0. \tag{5.19c}$$

Let us briefly examine the implications of eqs. (5.19). First note that in all cases $|\Gamma_G \Gamma_L| \leqslant 1$, and that a $|\Gamma_G \Gamma_L| = 1$ implies that R_G and R_L each are either 0 or ∞. Such extremes, however, imply loss-less sources and loss-less lines, hence are not practical. For this reason, we focus attention on situations where $|\Gamma_G \Gamma_L| < 1$. We can see that in such cases $V_L(t)$ will approach $V_0 R_L / (R_G + R_L)$ as N, and hence t, approach infinity; that is, V_L will settle to the value determined by dc considerations. Transients computed from eqs. (5.19) are illustrated in Figure 5.13b with R_G and R_L chosen as $R_G = 0.2 Z_0$ and $R_L = 4 Z_0$, that is, $\Gamma_G = -\frac{2}{3}$ and $\Gamma_L = \frac{3}{5}$.

Equations (5.19), and similar equations for V_G, provide the transients for arbitrary resistive terminations. Another method for finding the transients is the *graphical method*, which is discussed in Section 5.4 and which is also applicable to nonlinear resistive terminations. The accuracy of the graphical method is usually limited to a few percents by the limited precision in drawing lines: such limited accuracy could be detrimental in analog systems; in digital systems, however, it is readily tolerated.

5.2.5 Initial Conditions

The discussion of transients thus far has assumed that all voltages and currents are initially zero, which is rarely true in a digital system. However, the results are applicable to nonzero initial conditions by a simple addition (superposition) of the initial conditions.

Example 5.6 Figure 5.14a shows two emitter-coupled logic circuits connected by a transmission line that has a characteristic impedance of $Z_0 = 100 \ \Omega$ and a propagation delay of T_0 (connections of the ECL circuits to the power supply and to ground are not shown). The line is terminated by a resistor $R_L = Z_0 = 100 \ \Omega$ to a power supply voltage of -2 V; all loading by the second ECL circuit is neglected. The output of the first emitter-coupled logic circuit changes from -1.6 V to -0.8 V at time $t = 0$. Thus, initially all voltages are at -1.6 V, and the current in the line is $[-1.6$ V $- (-2V)] / 100 \ \Omega = 4$ mA. The resulting transients are obtained by adding these initial values to the transients of Figure 5.7b with $V_0 = 0.8$ V and $Z_0 = 100 \ \Omega$. The resulting transients at points G and L are shown in Figure 5.14b.

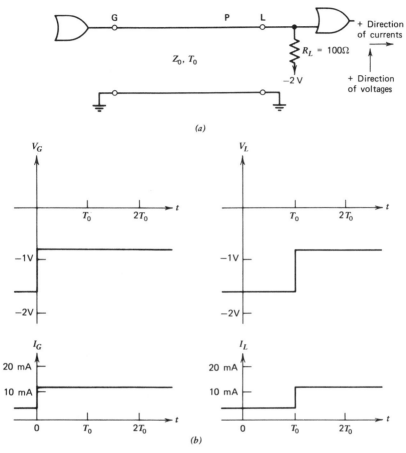

Figure 5.14 Transmission line driven by an emitter-coupled logic circuit. (*a*) The circuit; (*b*) transient response for a step voltage input.

5.3 CAPACITIVE TERMINATION

In the preceding discussion we assumed that each end of the transmission line was terminated by a resistance, and thus the boundary conditions could be described by Ohm's law—in some limiting cases with zero or infinite resistance. Ohm's law, however, becomes inadequate when a termination is not a pure resistance but, as is the case in a typical digital circuit, it also has a reactive component. This section discusses transients in transmission lines for capacitive termination, which is the most common reactive termination in high speed digital circuits.

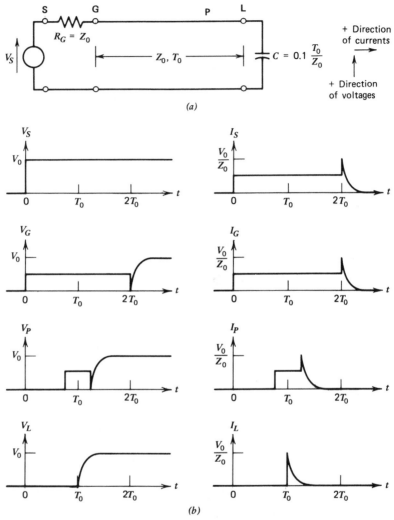

(a)

(b)

Figure 5.15 Transmission line terminated by a capacitance of $C = 0.1T_0/Z_0$. (a) The circuit; (b) transients for a step voltage input.

5.3.1 Transient Response

Two simple capacitive terminations are shown in Figures 5.15a and 5.16a. Figure 5.15a is obtained from Figure 5.9a by the addition of capacitance C; Figure 5.16a is similarly obtained from Figure 5.7a. Transients in these two circuits are described in the following discussion.

In the circuit of Figure 5.15a a transmission line is driven on its left end with a source resistance $R_G = Z_0$ and is terminated on its right end by a capacitance C. The resulting transients are illustrated in Figure 5.15b for a step voltage input V_0 and for a capacitance chosen as $C = 0.1T_0/Z_0$. These transients were found as follows:

The step voltage input V_0 shown in the top left graph of Figure 5.15b results in an immediate step voltage of $V_0/2$ in V_G, since series resistance R_G between points **S** and **G** in Figure 5.15a has a value of $R_G = Z_0$. This signal starts traveling to the right, reaches intermediate point **P** at time $t = 0.75T_0$, and the right end of the transmission line, point **L**, at time $t = T_0$. The capacitive load permits no instantaneous change in voltage V_L, thus V_L will remain zero for times immediately following $t = T_0$. For long times, however, the capacitance acts as an open circuit, hence according to Figure 5.9b a voltage of $V_L = V_0$ will be eventually established. The transition between the two voltage levels is an exponential with a timeconstant that is the product of capacitance C and the resistance it sees, which is Z_0. Hence, for $t \geqslant T_0$

$$V_L = V_0 \left[1 - e^{-(t-T_0)/(Z_0 C)}\right]. \tag{5.20}$$

For the specific choice of $C = 0.1\, T_0/Z_0$, we get

$$V_L = V_0 \left[1 - e^{-(t-T_0)/(0.1\, T_0)}\right], \tag{5.21}$$

which is shown in the bottom left graph of Figure 5.15b. Now, with

$$V_L = v_i + v_r, \tag{5.22}$$

with $v_i = V_0/2$, and with V_L from eq. (5.21) we get for v_r at point **L**:

$$v_r = V_L - v_i = V_0 \left[\tfrac{1}{2} - e^{-(t-T_0)/(0.1\, T_0)}\right]. \tag{5.23}$$

The reflected voltage v_r is thus $-V_0/2$ for times immediately following T_0, and it approaches $V_0/2$ with a timeconstant of $0.1T_0$. This reflected voltage reaches intermediate point **P** at time $t = 1.25T_0$ where it is added to the already present $V_0/2$, resulting in

$$V_P = \frac{V_0}{2} + V_0 \left[\tfrac{1}{2} - e^{-(t-1.25\, T_0)/(0.1\, T_0)}\right] = V_0 \left[1 - e^{-(t-1.25\, T_0)/(0.1\, T_0)}\right]$$

$$\tag{5.24}$$

for $t > 1.25T_0$, as shown in the graph for V_P in Figure 5.15. Similarly, for $t > 2T_0$ voltage V_G becomes

$$V_G = V_0 \left[1 - e^{-(t-2\, T_0)/(0.1\, T_0)}\right], \tag{5.25}$$

and no further reflections take place. The currents shown in Figure 5.15b are obtained by utilizing $v_i/i_i = Z_0$ and $v_r/i_r = -Z_0$ for the signals traveling to the right and to the left, respectively. An application of the results of Figure 5.15 is shown in the example that follows.

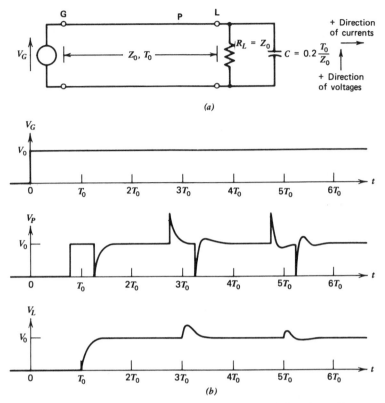

Figure 5.16 Transmission line terminated by its characteristic impedance Z_0 in parallel with a capacitance of $C = 0.2T_0/Z_0$. (a) The circuit; (b) transients for a step voltage input.

Example 5.7 The output of an emitter-coupled logic circuit is connected to the input of another one by a coaxial cable that has a characteristic impedance of $Z_0 = 50 \ \Omega$ and a delay of $T_0 = 2.5$ nsec. The cable is driven with a resistance of $R_L = Z_0 = 50 \ \Omega$ in series with its sending (generator) end as in Figure 5.15a. The ECL circuit loading the cable is represented by a capacitance of $C = 5$ pF. Thus, $T_0/Z_0 = 2.5$ nsec/50 Ω = 50 pF, hence $C = 5$ pF = $0.1T_0/Z_0$, and Figure 5.15b is applicable with $T_0 = 2.5$ nsec.

Note the characteristics of Figure 5.15b showing transients for capacitive termination as compared with Figure 5.9b (open circuit termination) and with Figure 5.11b (short circuit termination). Long-time changes in Figure 5.15b are identical with those of Figure 5.9b since for long times capacitance C acts as an open circuit. However, instantaneous changes in Figure 5.15b are identical with those of Figure 5.11b since for fast changes capacitance C acts as a short circuit.

5.3.2 Multiple Reflections

The addition of capacitance C in Figure 5.15a altered the transients but it did not lead to multiple reflections since one end of the transmission line remained terminated by Z_0. Figure 5.16a shows a circuit where a capacitance is added in parallel to the load resistance of Figure 5.7a: now the transmission line is not terminated by Z_0 on either end, hence multiple reflections will take place. These are shown in Figure 5.16b for the case when capacitance C is arbitrarily chosen as $C = 0.2T_0/Z_0$.[†]

From Figure 5.16b it would seem that the circuit of Figure 5.16a is impractical because of the sequence of sharp spikes in V_P of Figure 5.16b. We must

[†]The derivation of these transients, as well as those in Figure 5.17, are beyond the scope of this book. They are based on the following Laplace transform pairs:

$$\mathcal{L}^{-1}\left\{\frac{1}{(s+a)^n}\right\} = \frac{t^n e^{-at}}{(n-1)!}$$

and

$$\mathcal{L}^{-1}\left\{\frac{a^{n+1}}{s(s+a)^{n+1}}\right\} = 1 - e^{-at}\left[1 + at + \frac{(at)^2}{2!} + \cdots + \frac{(at)^n}{n!}\right].$$

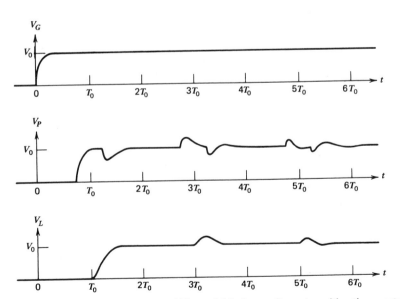

Figure 5.17 Transients in the circuit of Figure 5.16a for a voltage step with a timeconstant of 0.1T_0.

remember, however, that the voltage step with zero risetime shown in the up-most graph of Figure 5.16b for V_G is unrealistic in as much as a real signal would have a finite risetime. The effect of a finite risetime of V_G is illustrated (again without derivation) in Figure 5.17 where $V_G = V_0 \left[1 - e^{-t/(0.1 T_0)}\right]$. We can see that the reflections of Figure 5.17 are much smoother than those of Figure 5.16b and would be probably tolerable in a digital system. The results of Figure 5.17 are applied in the example that follows.

> **Example 5.8**   The output of an emitter-coupled logic circuit is connected to the input of another one by a coaxial cable with $Z_0 = 100\ \Omega$ and $T_0 =$ 2.5 nsec (see Figure 5.14a). The cable is driven at its left end by the output of the first ECL circuit, which is approximated as a zero-impedance source. The cable is terminated at its right end by a resistance of $R_L = Z_0 =$ 100 Ω in parallel with the input of the second ECL circuit, which is approximated by a capacitance of $C = 5$ pF. Thus, $0.2 T_0/Z_0 = 0.2 \times 2.5$ nsec/100 $\Omega = 5$ pF $= C$, and the transients of Figure 5.16b are applicable with $T_0 = 2.5$ nsec when the output of the first ECL circuit is a step voltage. The transients of Figure 5.17 are applicable, again with $T_0 = 2.5$ nsec, when the output of the first ECL circuit has a timeconstant of $0.1 T_0 = 0.25$ nsec.

5.4   NONLINEAR RESISTIVE TERMINATION

As an introduction to the treatment of a transmission line terminated by a non-linear resistance we first examine a linear case. Figure 5.18a shows an emitter-coupled logic circuit driving a transmission line that has a characteristic imped-ance of $Z_0 = 50\ \Omega$ and that is terminated at its right end by a resistor of $R_L =$ 200 Ω. (Connections of the ECL circuit to the power supply and ground are not shown). The output of the emitter-coupled logic circuit shown in Figure 5.18a is approximated in Figure 5.18b by a voltage source V_S in series with a resistance of $R_G = 10\ \Omega$. We examine the transients in the case when V_S changes from a logic 0 level of -1.6 V to a logic 1 level of -0.8 V at time $t = 0$.

Voltages V_G and V_L and currents I_G and I_L are shown as functions of time in Figure 5.18c; these are obtainable by the methods of the preceding section (see Figure 5.13b). The same information is given in Figure 5.18d, but in the *voltage-current plane*: This representation shows voltages vertically and currents horizontally. Load resistance R_L connected to -2 V is characterized by the voltage-current line marked "V_L versus I_L." With the directions of currents and voltages shown in Figure 5.18b, the combination of V_S and R_G is charac-terized for logic 0 and logic 1 outputs, respectively, by the "V_G versus I_G for logic 0 output" and by the "V_G versus I_G for logic 1 output" lines.

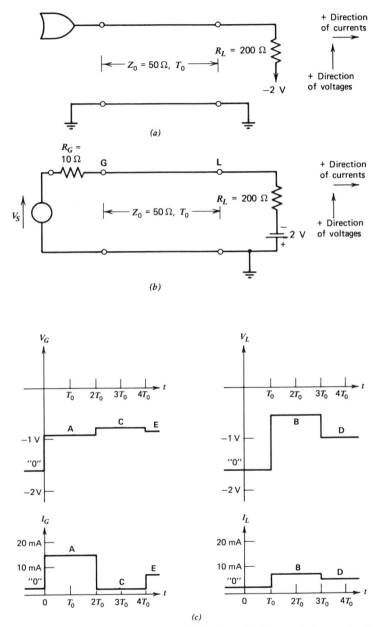

+ Direction
of currents

+ Direction
of voltages

$R_L = 200 \, \Omega$

$Z_0 = 50 \, \Omega, \, T_0$

−2 V

(a)

$R_G = 10 \, \Omega$

G

L

$R_L = 200 \, \Omega$

$Z_0 = 50 \, \Omega, \, T_0$

V_S

2 V

+ Direction
of currents

+ Direction
of voltages

(b)

V_G

$T_0 \quad 2T_0 \quad 3T_0 \quad 4T_0$

−1 V

A

C

E

"0"

−2V

V_L

$T_0 \quad 2T_0 \quad 3T_0 \quad 4T_0$

−1 V

B

D

"0"

−2 V

I_G

20 mA

A

10 mA

C

E

"0"

$0 \quad T_0 \quad 2T_0 \quad 3T_0 \quad 4T_0$

I_L

20 mA

10 mA

B

D

"0"

$0 \quad T_0 \quad 2T_0 \quad 3T_0 \quad 4T_0$

(c)

Figure 5.18 Multiple reflections in a transmission line with linear resistive terminations. (a) The circuit; (b) equivalent circuit; (c) voltage and current transients as functions of time for the transition from the logic 0 to the logic 1 state; (d) transients of V_G, V_L, I_G, and I_L shown in the voltage-current plane. (See page 166 for Figure 5.18d.)

165

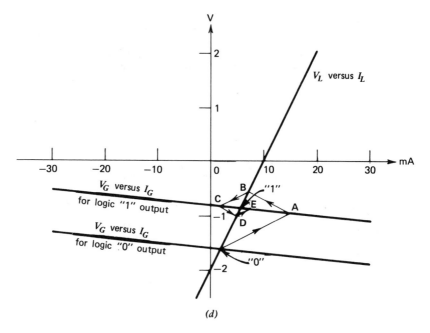

(d)

Figure 5.18 (continued)

Under static operating conditions, voltage V_G equals voltage V_L and current I_G equals current I_L. Thus, the static operating point for logic 0 output is given by the intersection of the "V_L versus I_L" line with the "V_G versus I_G for logic 0 output" line: this is marked as point "0" in Figure 5.18d. Similarly, the static operating point for logic 1 output is given by the intersection of the "V_L versus I_L" line with the "V_G versus I_G for logic 1 output" line, marked as point "1" in Figure 5.18d.

Transition from point "0" to point "1" takes place along the path marked "0," A, B, C, D, E, . . . , "1." These points are also identified in Figure 5.18c; point "0" appears in signals V_G, I_G, V_L, and I_L; points A, C, and E appear only in signals V_G and I_G; and points B and D appear only in signals V_L and I_L. The values of V_G and V_L in Figure 5.18c equal the projections of the corresponding points on the vertical axis in Figure 5.18d, while the values of I_G and I_L equal the projections of the corresponding points on the horizontal axis.

Note the pattern of the lines connecting points "0," A, B, C, D, E, . . . , "1" in Figure 5.18d. The line connecting point "0" and point A has a slope of 50 Ω, which equals Z_0 of the transmission line; lines between points B and C and between D and E have the same slopes. The line connecting points A and B has a slope of -50 Ω, which equals $-Z_0$ of the transmission line; the line between points **C** and **D** has the same slope. The recognition of this pattern in the interconnecting lines forms the basis of the *graphical method* of finding transients in a transmission line, which method is also applicable to nonlinear resistive terminations.

5.4.1 The Graphical Method

The graphical method of finding the transients in a transmission line terminated by nonlinear resistances originates from Bergeron.* Its use is illustrated here on a transmission line interconnecting two TTL inverters as shown in Figure 5.19a.

In Figure 5.19a, the output of a TTL inverter is connected to the input of a second one by a transmission line with a characteristic impedance of $Z_0 = 50$ Ω and a propagation delay of T_0. Typical voltage-current characteristics of the TTL inverters are shown in Figure 5.19b. When the output of the first inverter is logic 0, the static operating point is given by the intersection of the "V_{out} versus I_{out} for logic 0 output" curve with the "V_{in} versus I_{in}" curve, and is marked as point "0." Similarly, when the output of the first inverter is logic 1, the static operating point is given by the intersection of the "V_{out} versus I_{out} for logic 1 output" curve with the "V_{in} versus I_{in}" curve, and is marked as point "1."

We first find the transients for the case when the output of the first inverter changes from logic 0 to logic 1 at time $t = 0$. For times $t < 0$, V_{out}, V_{in}, I_{out}, and I_{in} are given by point "0" in Figure 5.19b. At time $t = 0$, the output of the first

*M. L. Bergeron, "Propagation D'Ondes Le Long Des Lignes Electriques," *Bulletin de la Societe Francaise des Electriciens*," 7, 979–1004 (October 1937). See also M. Abdel-Latif and M. J. O. Strutt, "Simple graphical method to determine line reflections between high-speed-logic integrated circuits," *Electronics Letters*, 4, 496–498 (November 1968).

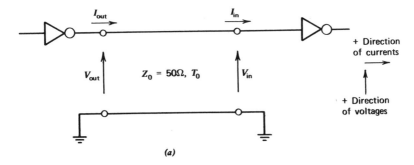

(a)

Figure 5.19 Multiple reflections in a transmission line between two TTL inverters. (a) The circuit; (b) typical static characteristics of the TTL inverters; (c) transition from logic 0 to logic 1 in the voltage-current plane; (d) V_{out}, I_{out}, V_{in}, and I_{in} as functions of time for a transition from logic 0 to logic 1; (e) transition from logic 1 to logic 0 in the voltage-current plane; (f) V_{out}, I_{out}, V_{in}, and I_{in} as functions of time for a transition from logic 1 to logic 0. (See page 168 for Figure 5.19b, page 169 for Figure 5.19c, page 170 for Figure 5.19d, page 171 for Figure 5.19e, and page 172 for Figure 5.19f.)

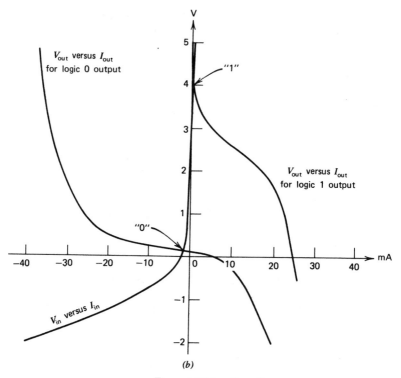

(b)

Figure 5.19 (continued)

inverter changes to logic 1 and as a result V_{out} and I_{out} instantaneously move from point "0" to *somewhere* onto the "V_{out} versus I_{out} for logic 1 output" curve. Hence, V_{out} and I_{out} will each *change* by some amount: we designate these changes by ΔV_{out} and ΔI_{out}, respectively. The output of the first inverter is connected to the left end of the transmission line, thus a signal characterized by ΔV_{out} and ΔI_{out} starts traveling to the right. However, for a signal traveling to the right the ratio of the change in voltage to the change in current is Z_0, that is, $\Delta V_{out}/\Delta I_{out} = Z_0$. This relation now constrains the new point of V_{out} and I_{out} on the "V_{out} versus I_{out} for logic 1 output" curve to be in a direction with respect to point "0" such that $\Delta V_{out}/\Delta I_{out} = Z_0 = 50\ \Omega$. This constraint is satisfied by a straight line with a slope of $50\ \Omega$ drawn through point "0." Thus, the intersection of this straight line with the "V_{out} versus I_{out} for logic 1 output" curve, shown as point **A** in Figure 5.19c, determines V_{out} and I_{out} for times $0 < t < 2T_0$.

The signal that is traveling to the right and that is characterized by a change from point "0" to point **A** in Figure 5.19c reaches the right end of the trans-

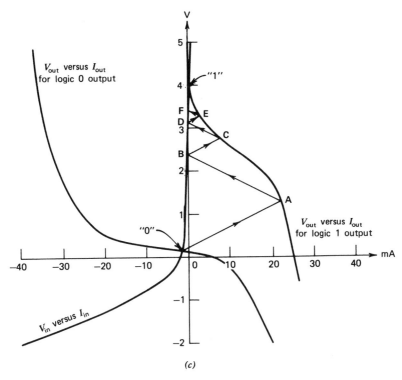

(c)

Figure 5.19 (continued)

mission line at time $t = T_0$. As a result, at time $t = T_0$, V_{in} and I_{in} will move from point "0" to another point somewhere onto the "V_{in} versus I_{in}" curve. V_{in} and I_{in} at this point, which we denote point **B**, are each composed of three constituents: the initial conditions given by point "0," the incident signal given by ΔV_{out} and ΔI_{out}, and a reflected signal which we characterize by ΔV_{in} and ΔI_{in}. The sum of the initial conditions given by point "0" and of the incident signal given by ΔV_{out} and ΔI_{out} is represented by point **A**. Thus, the difference between point **B** and point **A** represents the reflected signal, which is characterized by $\Delta V_{in}/\Delta I_{in} = -Z_0 = -50 \ \Omega$. Hence, point **B** can be found as the intersection of the "V_{in} versus I_{in}" curve with a straight line that originates from point **A** and that has a slope of $-50 \ \Omega$ (see Figure 5.19c).

The reflected signal characterized by ΔV_{in} and ΔI_{in} reaches the left end of the transmission line at time $t = 2T_0$ and will change V_{out} and I_{out} from point **B** to some other point on the "V_{out} versus I_{out} for logic 1 output" curve: we designate this new point as point **C**. By using a reasoning similar to that above, we can show that point **C** can be found as the intersection of the "V_{out} versus

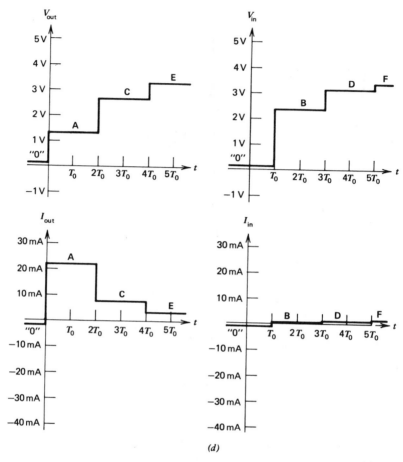

Figure 5.19 (continued)

I_{out} for logic 1 output" curve with a straight line that originates from point **B** and that has a slope of 50 Ω. The process may also be continued to obtain additional points, as shown in Figure 5.19c for points **D**, **E**, and **F**. Figure 5.19c also indicates that as the number of points is increased, they approach the final static operating point, point "1," as expected.

The diagram of 5.19c contains all information required for plotting V_{out}, I_{out}, V_{in}, and I_{in} as functions of time. The output of the first TTL inverter changes from logic 0 to logic 1 at time $t = 0$, hence V_{out} and I_{out} change from their initial values given by point "0" to the values given by point **A** at time $t = 0$. The projection of point "0" onto the vertical axis provides the initial value of $V_{\text{out}}(t < 0) = 0.15$ V, and the projection of point "0" onto the horizontal axis

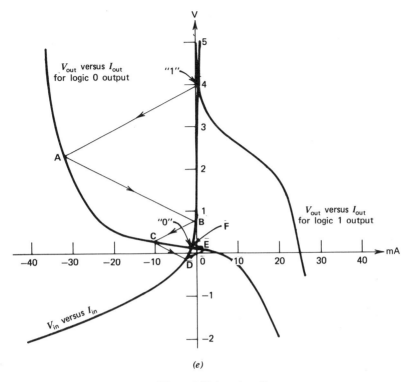

(e)

Figure 5.19 (continued)

provides the initial value of $I_{out}(t < 0) = -1.5$ mA. For times $0 < t < 2T_0$, the projection of point **A** onto the vertical axis provides V_{out} $(0 < t < 2T_0) = 1.3$ V, and the projection of point **A** onto the horizontal axis provides $I_{out}(0 < t < 2T_0) = 22$ mA. Further values of V_{out} and I_{out} can be similarly found from points **C** and **E** as shown in the left two graphs of Figure 5.19d.

At the right end of the transmission line, V_{in} and I_{in} remain at their initial values given by point "0" until time $t = T_0$, that is, $V_{in}(t < T_0) = 0.15$ V and $I_{in}(t < T_0) = -1.5$ mA. For times $T_0 < t < 3T_0$, the values of V_{in} and I_{in} are given by the projections of point **B** as $V_{in}(T_0 < t < 3T_0) = 2.35$ V and $I_{in}(T_0 < t < 3T_0) = 0.5$ mA. Further values of V_{in} and I_{in} are obtained from points **D** and **F** and are shown in the right two graphs of Figure 5.19d.

The transients in the circuit of Figure 5.19a were determined in Figures 5.19c and 5.19d for the case when the output of the first TTL inverter changes from logic 0 to logic 1. The transients can be found in a similar manner for a change from logic 1 to logic 0: these are shown in Figures 5.19e and 5.19f.

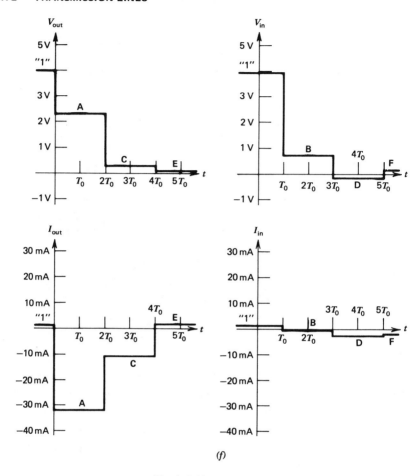

(f)

Figure 5.19 (continued)

5.4.2 Applications

Figures 5.19d and 5.19f show the transients in the circuit of Figure 5.19a for the case when the static characteristics of the TTL inverters are given by Figure 5.19b. These characteristics, however, are only typical, and thus are subject to variations from unit to unit. In a real application we have to consider the effects of these variations on the overall performance of the circuit.

The right top graph in Figure 5.19d provides the input voltage to the second TTL inverter V_{in} as a function of time for a change in the output of the first inverter from a logic 0 to a logic 1. At time $t = T_0$, V_{in} changes from 0.15 V to 2.35 V. Since 2.35 V is above the minimum 2.0 V required for safe recognition of a logic 1 by the second inverter, we might conclude that the output of the second inverter will indeed switch from logic 1 to logic 0 at time $T_0 + t_{pd}$, where t_{pd} is the propagation delay of the second inverter. Unfortunately, the voltage-current characteristics of the TTL inverters given in Figure 5.19b are only *typical*, that is, it is possible (at least in principle) to find *some* units with such characteristics; the *worst-case* characteristics may be much less favorable. In particular, we can see in Figure 5.19c that voltage V_{in} given by point **B** is strongly dependent on the location of point **A**. A realistic appraisal of the worst-case location of the "V_{out} versus I_{out} for logic 1 output" curve shows that point **A** can be significantly to the left of what is shown in Figure 5.19c resulting in a much lower location of point B and hence in a much lower V_{in} for times $T_0 < t < 3T_0$. Consequently, the second inverter might not recognize a logic 1 input until point **D** is reached at time $t = 3T_0$. Thus for some units an additional delay of $2T_0$ may be introduced, which can be quite detrimental to the overall performance.

Now we look at the case when the output of the first inverter changes from logic 1 to logic 0. The input voltage to the second inverter V_{in} is given in the right top graph of Figure 5.19f. The situation is marginal here too: for times $T_0 < t < 3T_0$ point **B** provides a $V_{in}(T_0 < t < 3T_0) = 0.75$ V, which is barely below the maximum permitted 0.8 V. Further, the location of point **B** in Figure 5.19e is strongly dependent on the location of point **A**, which may be significantly to the right of where it is shown in Figure 5.19e. Hence, the change to logic 0 might not be recognized by the second inverter until at least point **D** is reached at time $t = 3T_0$.

In order to arrive at a safe design, the worst-case voltage-current characteristics should be used instead of the typical ones shown in Figure 5.19b. Worst-case characteristics are, however, often difficult to assess. We are interested in properties in the vicinities of points **A** in Figures 5.19c and 5.19e, and characteristics of TTL circuits are commonly not specified there. Some information may be gleaned from the open-circuit logic 0 output specification of $V_{out} \leqslant$ 0.4 V at $I_{out} = -16$ mA and from the short-circuit logic 1 output specification of $I_{out} \geqslant 18$ mA at $V_{out} = 0$ V (specifications on propagation delays do not provide additional information). The worst-case characteristics of V_{out} versus I_{out} thus inferred are shown in Figure 5.20 by broken line. Also shown are the worst-case input characteristics (by broken line), and the typical characteristics copied from Figure 5.19b (by solid line).

The use of the worst-case characteristics results in additional delays in the

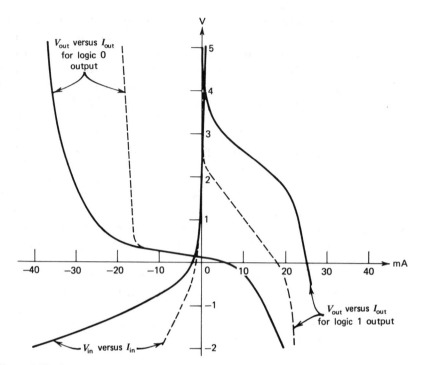

Figure 5.20 Static characteristics of a TTL inverter circuit. Typical characteristics are shown by solid lines, worst-case characteristics by broken lines.

recognition of logic level changes at the input of the second inverter in Figure 5.19*a*. The situation can be alleviated by using a transmission line with a higher characteristic impedance—coaxial cables with higher Z_0, however, are often also more bulky. The evaluation of the improvement resulting from higher-impedance transmission lines is left to the reader for exercise (see Problems 17, 18, and 19 at the end of the chapter).

†5.5 LOSSES

Thus far this chapter has assumed *ideal transmission lines* characterized by signals traveling without attenuation and distortion. In reality, a signal traveling in a transmission line is attenuated and distorted by several effects. These include dc resistance of the conductors which also increases for fast signals, leading to increased delay and risetime. Dc leakage of the dielectric is usually negligible in digital systems, but other effects such as the finite transverse size of the transmission line may result in significant distortion of fast signals.

5.5.1 DC Resistance

A signal traveling in a transmission line is attenuated by the series resistance of the conductors. The resistance R of a conductor with a length l, cross-sectional area A, and resistivity ρ can be written as

$$R = \rho \frac{l}{A},\qquad(5.26)$$

where for copper at room temperature $\rho \approx 1.7 \times 10^{-8}$ Ω m. In coaxial cables the dc resistance of the outer conductor is usually negligible with respect to the resistance of the inner conductor, thus eq. (5.26) need be evaluated only for the inner conductor.

Computation of the effects of dc resistance may be simplified by a crude approximation—which is, however, usually adequate in digital electronics—by which the total series dc resistance of the line is considered in one lump at one end of the line.

Example 5.9 The Type RG 174/U coaxial cable has an overall outer diameter of 0.105 in. ≈ 2.65 mm, a dielectric outer diameter of $d_2 = 0.06$ in. ≈ 1.5 mm, a copper inner conductor with a diameter of $d_1 \approx 0.016$ in. ≈ 0.4 mm, and a $Z_0 = 50$ Ω. Thus, the inner conductor has a cross-sectional area of

$$A = \pi (d_1/2)^2 = \pi\, 0.2^2 \text{mm}^2 = 0.126 \text{ mm}^2,$$

and the resistance of a 1-meter length is

$$R = 1.7 \times 10^{-8} \ \Omega \text{ m} \frac{1 \text{ m}}{0.126 \times 10^{-6} \text{ m}^2} = 0.135 \ \Omega.$$

A length of 3 m ($\approx$10 feet) of this cable is used as an interconnection between two ECL circuits. The resistance of this cable is 3 m $\times$ 0.135 Ω/m $\approx$ 0.4 Ω. The cable is terminated at its load end by a 50-Ω resistor. For simplicity we consider the cable as an ideal transmission line in series with a 0.4-Ω resistor at its load end. This results in an attenuation of 0.4 Ω/(50 Ω + 0.4 Ω) $\approx$ 0.008 = 0.8% and in a reflection coefficient of $\approx$0.4%, both negligible in a digital system.

5.5.2 Skin Effect

The foregoing considerations assumed that the full cross-sectional area of the center conductor is available for signal transmission—which is certainly correct for slowly varying signals. Fast changing signals, however, "crowd" near the surface of the conductor ("*skin effect*") resulting in signal distortion. For a step voltage input V_{in} that is zero for times $t < 0$ and that is $V_{in} = V_0$ for times $t > 0$, the output voltage of a transmission line terminated by its Z_0 can be written as

$$V_{out}(t) = V_0 \left(1 - \text{erf} \sqrt{\frac{\tau}{t - T_0}}\right), \tag{5.27}$$

illustrated in Figure 5.21.* For a coaxial cable, timeconstant τ is given as

$$\tau = \frac{\rho \mu l^2}{16\pi^2 Z_0^2 d_1^2}, \tag{5.28}$$

where for copper at room temperature $\rho \approx 1.7 \times 10^{-8}$ Ω m, $\mu = 4\pi \times 10^{-7}$ Ω sec/m, Z_0 is the characteristic impedance in ohms and l and d_1 (in identical units) are the length and the diameter of the inner conductor, respectively.[†]

Note that the transient in Figure 5.21 rises from zero to 10% in 0.75τ, to 50% in 4τ, but it takes approximately 120τ to reach the 90% point. Further, the 95% point (not shown in Figure 5.21) is reached in about 500τ.

*The function erf(x) is the *error function* defined as

$$\text{erf}(x) = \frac{2}{\sqrt{\pi}} \int_0^x e^{-z^2} \, dz.$$

[†]See, for example, A. Barna, *High-Speed Pulse Circuits*, Wiley-Interscience, New York, 1970, p. 66.

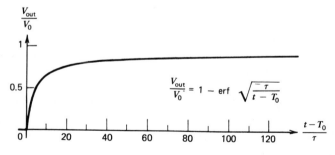

Figure 5.21 Normalized transient in a transmission line dominated by skin-effect losses.

Example 5.10 For the coaxial cable described in Example 5.9, $\rho = 1.7 \times 10^{-8}$ Ω m, $\mu = 4\pi \times 10^{-7}$ Ω sec/m, $l = 3$ m, $Z_0 = 50$ Ω, and $d_1 = 0.4$ mm $= 4 \times 10^{-4}$ m. Thus, from eq. (5.28),

$$\tau = \frac{1.7 \times 10^{-8} \ \Omega \ \text{m} \times 4\pi \times 10^{-7} \ (\Omega \ \text{sec/m}) \times 3^2 \, \text{m}^2}{16\pi^2 \ 50^2 \ \Omega^2 \ 4^2 \times 10^{-8} \ \text{m}^2} = 0.003 \ \text{nsec.}$$

Hence the delay of the 50% point is $\approx 4 \times 0.003$ nsec $= 0.012$ nsec and the delay of the 95% point is $\approx 500 \times 0.003 = 1.5$ nsec, both in addition to the $T_0 = 15$ nsec delay of the cable.

5.5.3 Additional Effects

Even though the dominant parasitic contributor to the delay and to the risetime is usually the skin effect, often the results obtained from eq. (5.27) are in error—sometimes by as much as a factor of 5 (either way!). Although time-domain measurements can be qualitatively correlated with frequency-domain measurements,* quantitative contributions of effects other than the skin effect have not been found. However, three additional effects have been identified as dielectric loss, losses in the braided outer conductor of a coaxial cable, and the appearance of additional propagation modes (modes other than TEM) in coaxial cables with diameters that approach the product of τ of eq. (5.28) and the speed of light.

PROBLEMS

1. Calculate the characteristic impedance, velocity of signal propagation, and capacitance per foot (≈ 30 cm) of a coaxial cable with $d_1 = 2.5$ mm (≈ 0.1 in.), $d_2 = 7.5$ mm (≈ 0.3 in.), and $\epsilon_r = 2.25$.

2. Find the characteristic impedance of a microstrip (Figure 5.5) on a printed-circuit board that has a thickness of 1.5 mm (≈ 0.06 in.) and an $\epsilon_r = 5$; the width of the printed strip is 2.5 mm (≈ 0.1 in.).

3. Compute the velocity of signal propagation and the capacitance per meter for the microstrip of Problem 2 above; use ϵ_{eff} instead of ϵ_r in eqs. (5.4) and (5.8).

4. Complete Figure 5.9c by sketching V_G, I_G, V_P, I_P, V_L, and I_L in the circuit of Figure 5.9a for an input pulse with a width of $3T_0$.

†5. The output of an emitter-coupled logic circuit is connected to the input of

---

*M. P. Ekstrom, "On the relationship of excess phase dispersion to the anomalies of broadband system compensation," Report UCRL-50654, Lawrence Radiation Laboratory, University of California, Livermore, June 20, 1969.

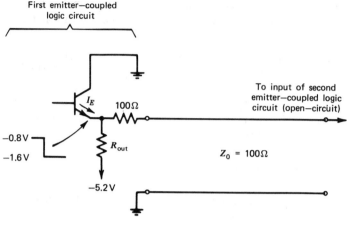

Figure 5.22

another one by a coaxial cable with a characteristic impedance of $Z_0 = 100$ Ω and a propagation delay of $T_0 = 50$ nsec. The interconnection can be represented by the circuit of Figure 5.9a, where the voltage source V_S now represents the output of the first emitter-coupled logic circuit while the input of the second emitter-coupled logic circuit is connected to point **L** and is approximated as an open circuit (see Figure 5.22). Consider the negative current pulse shown in the lower part of Figure 5.9c and find the maximum permitted value of the output pulldown resistor R_{out} of the first emitter-coupled logic circuit such that $I_E \geqslant 0$ at all times.

6. Check the correctness of each signal in Figure 5.11b.

7. Complete Figure 5.11c by sketching I_S, I_G, V_P, I_P, V_L, and I_L in the circuit of Figure 5.11a for an input pulse with a width of $3T_0$.

8. Identify the origin of each of eqs. (5.16a) through (5.16e).

†9. (*a*) Demonstrate that when $R_G > Z_0$ and $R_L > Z_0$, the transmission line of Figure 5.13a may be replaced by a capacitance. Specifically, show that when $R_G > Z_0$ and $R_L > Z_0$, for $t > 0$ eq. (5.19a) may be approximated as

$$V_L \approx V_0 \frac{R_L}{R_G + R_L} (1 - e^{-t/\tau}) \qquad (5.29a)$$

with

$$\tau = \frac{2T_0}{\ln [1/(\Gamma_G \Gamma_L)]}. \qquad (5.29b)$$

(b) Establish that limits on the error committed in the approximation of (a) above can be described as

$$V_0 \frac{R_L}{R_G + R_L} [1 - e^{-(t-T_0)/\tau}] \leqslant V_L \leqslant V_0 \frac{R_L}{R_G + R_L} [1 - e^{-(t+T_0)/\tau}],$$

that is, show that the graph of the exact $V_L(t)$ is always between the two graphs given by shifting eq. (5.29a) by $+T_0$ and $-T_0$.

(c) Show that when $R_G \gg Z_0$ and $R_L \gg Z_0$, then in eq. (5.29b)

$$\tau \approx \frac{R_G R_L}{R_G + R_L} C_0,$$

where

$$C_0 = \frac{T_0}{Z_0}.$$

†10. Derive current signals $I_S, I_G, I_P,$ and I_L shown in Figure 5.15b.

†11. Compare the methods of terminating a transmission line that interconnects two emitter-coupled logic circuits. In one method, shown in Figure 5.14a (also outlined in Figure 5.7a), the transmission line is terminated by a parallel load resistance $R_L = Z_0$ at its load end. In the other method, outlined in Figure 5.9a, the transmission line is terminated by a series generator resistance $R_G = Z_0$ at its generator end. Assume that the characteristic impedance of the transmission line, Z_0, is the same in both methods and compare the following properties:

(a) Maximum current swing required from the driver circuit by taking into account the negative I_S shown in Figure 5.9c.

(b) Standby power required in each of the two logic states, including the standby power needed to supply the negative I_S shown in Figure 5.9c.

(c) Possibility of tapping the line at an intermediate point for use as an input to another ECL circuit.

(d) Risetime and delay contributed by a capacitive load at the load end of the transmission line.

(e) Multiple reflections resulting from a capacitive load at the load end of the transmission line.

†12. Repeat Problem 11 (a) through (e) above, except assume a characteristic impedance for the first method (Figure 5.14a) that is twice as large as the characteristic impedance used in the second method (Figure 5.9a).

†13. Derive the signals shown in Figure 5.16b, but only for times $t < 3T_0$.

14. Find V_P and V_L for times $t \leqslant 3T_0$ in Figure 5.7a for a step voltage input V_G if an *inductance* of $L = 0.2T_0 Z_0$ is added in series with load resistance R_L.

15. Find the transients of $V_G, I_G, V_L,$ and I_L in the circuit of Figure 5.18a for

Z_0 = 100 Ω, that is, prepare graphs similar to those of Figure 5.18c but with Z_0 = 100 Ω.

16. Find the transients of V_{out}, I_{out}, V_{in}, and I_{in} in the circuit of Figure 5.19a for Z_0 = 100 Ω, that is, prepare graphs similar to those of Figures 5.19d and 5.19f but with Z_0 = 100 Ω. Use the typical voltage-current characteristics given in Figure 5.19b.

17. Use the worst-case voltage-current characteristics given by broken lines in Figure 5.20 and find the transients of V_{out}, I_{out}, V_{in}, and I_{in} in the circuit of Figure 5.19a.

18. Repeat Problem 16, but by using the worst-case voltage-current characteristics given by broken lines in Figure 5.20.

[†]19. Compare the transients of Figures 5.19d and 5.19f and of Problem 16 with those obtained in Problems 17 and 18.

[†]20. Look up a standard TTL inverter data sheet and demonstrate that the static characteristics shown in Figure 5.20 are reasonable.

[†]21. Find the 10 to 90% risetime resulting when a step voltage is passed through 30 m (≈100 feet) of coaxial cable that has a copper inner conductor with a diameter of d_1 = 0.09 in. ≈ 2.3 mm; Z_0 = 50 Ω. Also find the dc resistance of the center conductor.

REFERENCES

1. A. Barna and D. I. Porat, *Integrated Circuits in Digital Electronics*, Wiley-Interscience, New York, 1973.
2. R. E. Matik, *Transmission Lines for Digital and Communication Networks*, McGraw-Hill, New York, 1969.

ANSWERS TO SELECTED PROBLEMS

Chapter 2

8: 0.18 μH. 19: $(\ln 2)/(2\pi f_0)$, $(\ln 9)/(2\pi f_0)$, $1/(2\pi f_0)$, $1/(\sqrt{2\pi} f_0)$, 0.
[†]20: $M_0 (\ln 2)/(2\pi f_0 A_0)$, $M_0 (\ln 9)/(2\pi f_0 A_0)$, $M_0/(2\pi f_0 A_0)$, $M_0/(\sqrt{2\pi} f_0 A_0)$.
[†]21: (a) $t_{D_{\text{overall}}} = M_0 n/(2\pi f_0 A_0)$, $t_{R_{\text{overall}}} = M_0 \sqrt{n}/(\sqrt{2\pi} f_0 A_0)$.

Chapter 3

1: 0.0098 nA, 0.009997 nA. 3: 2.5 Ω, 40 pF. 7: 85 psec.
18: t_d = 414 psec, t_r = 304 psec.

Chapter 4

2: 0.99, 19.8 mA, 0.198 mA. 4: 80 pF, 2 pC. 5: 0.125, 2.61.
[†]9: (a) 0.9 V $(1 - e^{-t/100 \text{ psec}})$, (b) 0.09 V $(1 - e^{-t/100 \text{ psec}})$.
12: (a) 232.5, (b) 208.6. 13: (a) 2013 pF, (b) 87 nH.

Chapter 5

1: 99 Ω, 20 cm/nsec, 15.2 pF/ft. 2: 57.7 Ω. 3: 17.3 cm/nsec, 100 pF/m. [†]5: 900 Ω.

INDEX